Advanced Logic Design Techniques

Zhang Dan

ABSTRACT

Because of the needs for producing highly reliable products and reducing product development time, Accelerated Life Testing (ALT) has been widely used in new product development as an alternative to traditional testing methods. The basic idea of ALT is to expose a limited number of test units of a product to harsher-than-normal operating conditions to expedite failures. Based on the failure time data collected in a short time period, an ALT model incorporating the underlying failure time distribution and life-stress relationship can be developed to predict the product reliability under the normal operating condition. However, ALT experiments often consume significant amount of energy due to the harsher-than-normal operating conditions created and controlled by the test equipment used in the experiments. This challenge may obstruct successful implementations of ALT in practice.

In this dissertation, a new ALT design methodology is developed to improve the reliability estimation precision and the efficiency of energy utilization in ALT. This methodology involves two types of ALT design procedures – the sequential optimization approach and the simultaneous optimization alternative with a fully integrated double-loop design architecture. Using the sequential optimum ALT design procedure, the statistical estimation precision of the ALT experiment will be improved first followed by energy minimization through the optimum design of controller for the test equipment. On the other hand, we can optimize the statistical estimation precision and energy consumption of an ALT plan simultaneously by solving a multi-objective optimization problem using a controlled elitist genetic algorithm. When implementing either of the methods, the resulting statistically and energy efficient ALT plan depends not only on the reliability of the product to be evaluated but also on the physical characteristics of the test equipment and its controller. Particularly, the statistical efficiency of each candidate ALT plan needs to be evaluated and the corresponding controller capable of providing the required stress loadings must be designed and simulated in order to evaluate the total energy consumption of the ALT plan. Moreover, the realistic physical constraints and tracking performance of the test equipment are also addressed in the proposed methods for improving the accuracy of test environment.

In this dissertation, mathematical formulations, computational algorithms and simulation tools are provided to handle such complex experimental design problems. To the best of our knowledge, this is the first methodological investigation on experimental design of statistically precise and energy efficient ALT. The new experimental design methodology is different from most of the previous work on planning ALT in that (1) the energy consumption of an ALT experiment, depending on both the designed stress loadings and controllers, cannot be expressed as a simple function of the related decision variables; (2) the associated optimum experimental design procedure involves tuning the parameters of the controller and evaluating the objective function via computer experiment (simulation). Our numerical examples demonstrate the effectiveness of the proposed methodology in improving the reliability estimation precision while minimizing the total energy consumption in ALT. The robustness of the sequential optimization method is also verified through sensitivity analysis.

CHAPTER 1

INTRODUCTION

1.1 Background and Motivation

Rapid growth of new products and global competition are placing an increasing demand on reliability testing to meet ever-increasing expectations for high reliability and substantial reduction in product development time. For highly reliable products, it is often difficult, if not impossible, to observe failures in a short time period under normal operating conditions. When failure data is scarce, precise reliability estimation via statistical modeling becomes quite difficult. To overcome this challenge, Accelerated Life Testing (ALT) has been widely used to estimate the reliability of a product by applying harsher-than-normal operating conditions that accelerate the interested failure process (Bagdonavičius and Nikulin, 2002; Nelson, 1990; Elsayed, 1996; Meeker and Escobar, 1998).

By statistically determining an ALT model consisting of a failure time distribution and the associated life-stress relationship based on a modest number of failure times, the product's long-term reliability under the normal operating conditions can be extrapolated with respect to both time and stresses. Nowadays, as technologies advance, reducing the uncertainty in such estimates becomes more important and challenging than ever before. The precision of reliability estimate depends on both the ALT model and the test plan adopted. An appropriate ALT model should fit sufficiently well to the ALT data in order to extrapolate the product's reliability. On the other hand, an optimum ALT plan determines the stress loadings of accelerating stresses (see Figure 1-1 for constant-stress, ramp-stress, cyclic-stress, etc.), the number of test units allocated to each stress level, and other experimental variables in ALT to improve the estimation precision. Without scientifically designed ALT plans, it is likely to obtain imprecise reliability estimates. This is crucial for most applications demanding high reliability, as indispensable ALT experiments are often time-consuming and imprecise reliability estimates will undoubtedly influence subsequent decisions on reliability improvement, warranty, and maintenance.

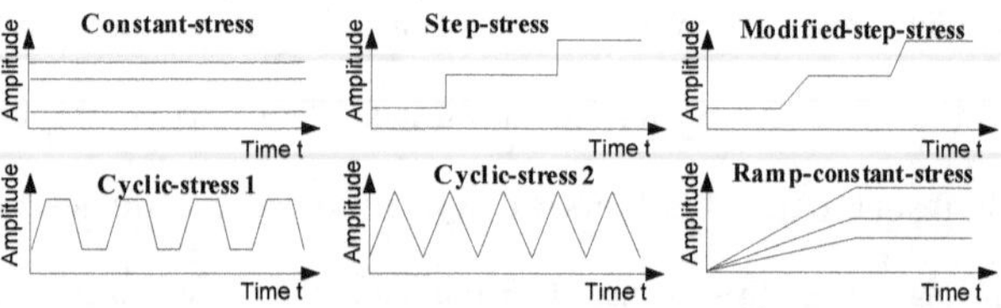

Figure 1-1: Various stress loadings widely used in ALT

A question often overlooked in planning ALT is concerned with the energy consumption of an ALT experiment. Indeed, different stress loadings used in ALT may exhibit significant differences in energy consumption. For example, Figure 1-2 shows three cyclic-stress loadings (thermal cycling, Z_1, Z_2, and Z_3) and their one-cycle power consumption values in horsepower (hp). The data were collected using a power meter on the 380 Volt power lines of a temperature/humidity test chamber used in an ALT experiment to study the reliability of solder joints. In many real-world applications, significant amounts of energy will be consumed in a series of ALT experiments due to the creation of harsher than normal test conditions. This often becomes an obstacle to successful implementations of ALT in practice due to budgetary and time constraints. As a result, it would be desired to minimize the energy consumption of ALT without increasing the sample size of test units and the length of test duration while still achieving the required precision of reliability estimate.

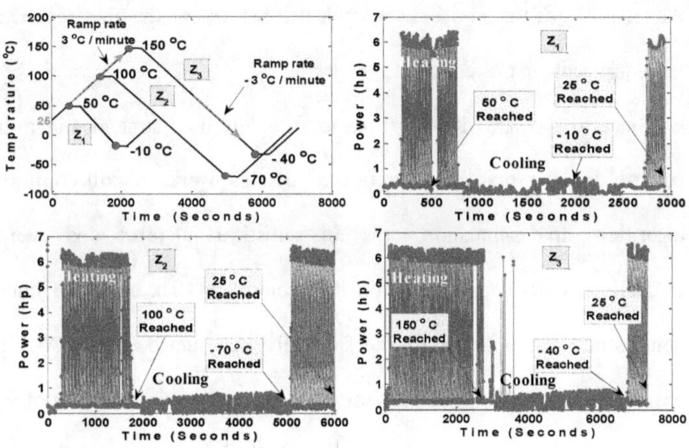

Figure 1-2: Cyclic-stress loadings vs. power consumption

1.2 Contributions

The goal of this dissertation is to develop a generic methodology for the optimum design of ALT plans for highly reliable products with respect to both statistical and energy efficiency. In order to achieve this goal, we first investigate the interrelationship among adopted test plans, test equipment control, and the associated energy consumption. In our investigation, we found functional relationship among the parameters of adopted test plan and control strategy, which is the foundation to realize energy reduction by manipulating these parameters. A new methodology is then developed to minimize the total energy consumption of an ALT experiment without sacrificing the experiment's statistical efficiency. This method provide flexibility in optimization, which allows us to optimize the statistical estimate accuracy and energy consumption sequentially or simultaneous. Therefore, the resulting statistically and energy efficient ALT plan depends not only on the reliability of the product to be evaluated but also on the physical characteristics of the test equipment and its controller.

Since PID control is one of the most widely used strategies in instrumentation and control, for the practical significance of this work, a collection of computational algorithms and simulation tools for statistical inference and energy consumption evaluation is developed based on PID controllers. The result of numerical examples demonstrates the potential of interfacing the optimum design of ALT plans and active energy reduction through statistical modeling and equipment control strategies. To the best of our knowledge, this is the first methodological investigation on experimental design of statistically and energy efficient ALT. Moreover, this is also the first approach considering the tracking performance of test equipment while designing optimum ALT plans. The new experimental design methodology is different from most of the previous work on planning ALT in that:

1) A first-time investigation on the energy consumption of an ALT experiment (somehow related to the cost of the experiment), depending on both the designed stress loadings and controllers, which cannot be expressed as a simple function of the related decision variables;

2) The associated sequential and simultaneous optimum experimental design procedures involves tuning the parameters of the controller and evaluating the objective function via computer experiment (simulation);

3) Consider the cases where the test equipment has or hasn't an adjustable controller;

4) Address the tracking performance of test equipment during optimum experimental design process;

5) Demonstrate the robustness of resulting ALT plans against the uncertainty in test equipment model.

1.3 Dissertation Organization

This dissertation is organized into six chapters. The first chapter introduces the background and motivation of this study. A thorough literature review of optimum ALT plans is provided in Chapter 2 to highlight the main contributions of this dissertation. Chapter 3 first introduces the traditional optimum ALT planning method and investigates the methods to assess the energy consumption of an ALT experiment. Then a new methodology for designing statistically and energy efficient ALT plans is delineated. Based on this methodology, the mathematical formulations for sequential optimization procedure are provided. This sequential optimization approach improves the statistical estimation accuracy first followed by energy minimization. Algorithms and simulation tools for solving such complex experimental design problems are also described in detail in this chapter. Chapter 4 presents a fully integrated multi-objective optimization framework for simultaneously optimizing the statistical and energy efficiency of an ALT experiment along with the corresponding algorithms and simulation tools. Numerical examples are provided in Chapter 5 to demonstrate the advantages of statistically and energy efficient ALT plans over traditional optimum ALT plans focusing only on statistical efficiency. Sensitivity analysis is also performed in this chapter to identify the robustness of sequentially optimum ALT plans against the uncertainty in test equipment model. Conclusions about the contributions of this research efforts, as well as future research directions are provided in Chapter 6.

CHAPTER 2

LITERATURE REVIEW

This chapter reviews the previous work on optimum design of ALT plans with respect to stress loading profiles and optimization criteria. The purpose of this literature review is not only to provide background information of current ALT planning methods, but also discuss the gap in current research, which will be addressed in this dissertation.

2.1 Optimum Design of ALT Plans

In 1962, Chernoff (1962) firstly investigated the optimal ALT designing method for exponential distribution to estimate the failure rate of a device at normal operation condition. The resulting ALT plans are regarded as local optimal due to their dependence on the true values of ALT model parameters. Since then, extensive research has been conducted on the optimum design of ALT plans regarding decision variables including stress loading profiles, unit allocation at each stress level, censoring time, etc.

As for constant-stress ALT, Nelson and Kielpinski (1976) obtained the optimum plans and the best traditional plans for estimating the medians of the normal and lognormal distributions, respectively. These plans are designed for testing with Type I censored data. Compared with the traditional plans, the optimum ALT plans are of better statistical efficiency while being less robust in general. Meeker and Hahn (1977) considered the optimal allocation of test units to accelerated stress conditions with the objective of minimizing the variance of product's log reliability estimate under the design stress condition. Nelson and Meeker (1978) provided the optimum test plans for Type-I censoring to estimate the percentiles of the Weibull and extreme-value distributions at a specified design stress. Nelson and Kielpinski (1976), Meeker and Hahn (1977) and Nelson and Meeker (1978) all indicate that more test units should be allocated at the lower stress level than at the higher one. Meeker (1984) compared the statistically optimum test plans with two stress levels to some more practical test plan involving three stress levels for Weibull and Log-normal Distributions. According to comparison results, although the compromise test plans generally provides less precision than the optimum ALT plans, they tend to be more robust to departures from model assumptions. Meeker and Hahn (1985) provided some practical guidelines for designing theoretically optimum and alternative ALT plans, which can be easily used by engineers. Nelson (1990) systematically reviewed the methods for designing statistically optimum and compromise ALT plans with single stress.

Since the work of Nelson (1984) and Boyko and Gerlach (1989) discovered that the scale parameter for a location-scale distribution should not be independent with stress changes, Meeter and Meeker (1994) extended the existing method for planning ALT with censored data to a non-constant scale parameter model which depend on stress and presented test plans for a large range of practical tests. Yang (1994) investigated a method for designing 4-level constant-stress optimal ALT plans considering different censoring times. Compared with the 3-leval test plans, the optimum 4-level ALT plans are more robust to the departures from modeling assumptions while requiring shorter test duration. In 2005, Nelson (2005) comprehensively summarized the research on ALT planning since the work done by Chernoff. Liao (2009) addressed a 3-level compromise constant-stress ALT plan for estimating the long-run cost rate of a periodical replacement policy when the normal operating condition is fixed. Recently, Yang (2010) considered 3-level best compromise ALT plans for predicting the warranty cost and its confidence interval for a product population that will experience varying stress levels in the field. Liu and Tang studied the optimum design of constant-stress ALT plans involving scheduled inspections (2013) by simultaneously optimize stress levels, sample allocation, and inspection times for log-location scale distributions.

Regarding step-stress ALT (SSALT), Miller and Nelson (1983) presented the cumulative exposure model and obtained the optimum simple step-stress test plans for time-step tests and failure-step tests. Both types of optimum plans aim to minimize the asymptotic variance of the maximum likelihood estimate (MLE) for the mean life of the exponential distribution at the design stress and have the same estimation precision as the corresponding optimum test for constant-stress. Bai et al. (1989) extended the results to the case where type I censoring is considered. Bai and Chung (1991) investigated similar optimum simple step-stress test plans for type I censoring with replacement and conducted sensitivity analysis with respect to the departures from the pre-estimated model parameters. Bai and Chun (1991) studied the optimum simple SSALTs for products with independent competing causes of failure. These plans have the minimal sum of asymptotic variances of log mean product lives over all failure causes. Chung and Bai (1998) considered the optimal design of SSALT to estimate a specified quantile at design stress in which a cumulative exposure model for the effect of changing stress is assumed. Khamis and Higgins (1996) presented optimum 3-step SSALT plans assuming a linear or quadratic life-stress relationship. They also proposed some compromise plans along with a criterion for compromise plan selection. Khamis (1997) extended the existing experimental design method for optimum SSALTs with only one accelerating variable to the case where M-level step-stress and k stress variables are considered. Xiong (1998) addressed the statistical inference of a simple SSALT model with Type II censoring, which can be applied on data with any sample size. Yeo and Tang (1999) considered planning multiple-step SSALT with a target acceleration factor using a sequential method. They found that compared with the optimal simple SSALT of the same target acceleration factors, the 3-step SSALT can ensure the valid life-stress relationship by slightly reducing estimation accuracy. Xiong and Milliken (1999) studied statistical models in SSALT with random stress change times and investigated corresponding optimum simple SSALT plans with exponential distribution. Alhadeed and Yang (2002) investigated the optimal simple SSALT design for Weibull distribution with unknown shape parameter using Khamis-Higgins model. Xiong and Ji (2004) studied the optimum design of a simple step-stress test plan involving grouped and censored data (as it gives a guide to experimenters about when the intermittent inspection and the stress change should be carried out during the test). Elsayed and Zhang (2007) also investigated a method for designing optimum simple SSALTs. Their method minimizes the asymptotic variance of reliability estimation via lower stress level and stress changing point selection. Gouno (2007) focused on temperature accelerated step-stress life testing assuming an Arrhenius life-stress relationship and addressed the problems of choosing the step duration and the stress levels. Moreover, Xu and Fei (2007) considered the optimum SSALT plans involving two independent stress variables and provided some guidelines for designing such tests. As an extension of previous work, Ma and Meeker (2008) developed a general method for designing statistically optimum ALT plans involving multi-step-stress profiles for log-location-scale Distributions. Fard and Li (2009) presented a design method for simple SSALTs using Khamis-Higgins model which aims to minimize the uncertainty in reliability estimation at censoring time via stress change time determination. Liu and Qiu (2011) studied planning of step-stress ALT with independent competing risks.

In previous ALT planning problem, the estimation precision of resulting optimum ALT plans usually depend on the selection of pre-estimated model parameters and the number of available test units. Unlike most work on ALT planning, Erkanli and Soyer (2000) developed a Bayesian decision approach to design m-level constant-stress ALT plans for exponential distribution using surface smoothing techniques. Zhang and Meeker (2006) studied Bayesian methods for planning constant-stress ALTs with censored data from log-location scale distributions. The developed optimum ALT plans should maximize the value of the adopted Bayesian criterion which is based on the estimation precision of a pre-specified quantile at design stress. Considering the situation when the number of available test units is limited, Ma and Meeker (2010) investigated a new simulation based method for searching three-level constant stress compromise ALTs with small sample sizes while providing acceptable estimation precision. Liu and Tang (2010) described a Bayesian method to design constant-stress ALTs for repairable systems with multiple independent failure modes. Yuan and Liu (2011) and Yuan et al. (2012) proposed Bayesian methods for planning simple and three-level SSALTs with Weibull distribution, which provide a way to handle the uncertainty in the planning values of model parameters.

In the literature, most work on the optimum design of ALT plans assumed instantaneous changes in stress levels, which may not be realistic in many cases. To handle time-varying stress levels with finite changing rates, Park and Yum (1998) developed a modified step-stress ALT plan and a modified constant-stress ALT plan considering limited stress ramp rates for exponentially distributed lifetimes. Liao and Elsayed (2010) studied an optimum ramp-stress ALT plan where the ramp rate is the only stress-related design variable in addition to the sample size. Hong et al. (2010) designed a new optimum ramp-stress ALT plan by simultaneously determining the ramp rate and lower start level of stress based on a cumulative exposure model.

Recently, Liao and Elsayed (2010) investigated the equivalency of ALT plans with different stress loadings. Two ALT plans are equivalent if they have the same statistical efficiency. General equivalent ALT plans and special types of equivalent ALT plans were studied for log-location-scale distributions. For multi-objective equivalent ALT plans when several estimation precision criteria, such as D-optimality and C-optimality are considered simultaneously, readers are referred to Liao and Li (2008).

With rare exceptions, the following objectives are often considered in planning statistically efficient ALT: (1) minimizing the asymptotic variance of the MLE of mean-time-to-failure or a failure time percentile, among others, under normal operating conditions, and (2) minimizing the asymptotic variance of the MLE of a model parameter or the determinant (D-optimality) of the variance-covariance matrix of model parameter estimates.

2.2 Optimization Criteria Comparison

In the literature, although extensive research has been conducted on the optimal design of ALT plans using two types of optimization criteria, respectively, little research has investigated the comparison of C-optimality and D-optimality with respect to their performance on reducing the uncertainty in estimation. Based on the existing literature, the characteristics of D-optimality and C-optimality can be summarized as follows:

1) D-optimality can effectively minimize the uncertainty of estimated model parameters (Gao et al., 2011; Ginebra and Sen, 1998; Xu and Fei, 2009) by minimizing the joint statistical-confidence region volume for the model parameters (Guo and Pan, 2007).

2) C-optimal ALT plans that minimize the asymptotic estimation variance at the usage condition among all test plans can provide more confidence in the resulting functional estimates (Yang and Pan, 2013). So C-optimality is widely used to minimize the asymptotic variance of the function of the estimated model parameters (Gao et al., 2011).

3) Although both D-optimality and C-optimality rely significantly on the validity of model assumption and the prior estimation of model parameters (Meeker and Escobar, 1998; Monroe, 2009; Nelson, 1990), C-optimal ALT plans are more robust than the ALT plans designed using D-optimality in the presence of small deviation from the assumed

model (Ginsburg and Ben-Gal, 2006; Monroe, 2009; Monroe etc., 2010).

4) For ALT plans with stress levels changed at a finite rate, such as modified step-stress, C-optimization criterion can provide stable results when the stress-increasing rate is not too large (Park and Yum, 1998). In addition, Escobar and Meeker (1995) suggested that the optimization criteria for designing an optimal ALT plan should be chosen according to the purpose of the experiment, which requires the researcher (or reliability engineers) to fully understand the testing objectives before designing an ALT experiment (Xu and Fei, 2009).

While most of the previous work on optimum ALT planning only deal with the optimization of statistical estimation precision, it is important to realize that energy consumption during experiments are related to non-negligible cost and should also be addressed in designing ALTs. In this dissertation, our goal is to develop a new ALT design methodology to improve the functional estimation precision of an ALT experiment while minimizing the experiment's total energy consumption. C-optimality is used to optimize the statistical efficiency of the desired ALT plan given its characteristics stated above. However, this criterion is only estimation-precision related, which cannot directly address the energy consumption of an ALT experiment. Then, a fundamental question to be raised is: "Can we design an efficient ALT experiment that consumes the minimal amount of energy without sacrificing the reliability estimation precision?" This question will be answered in the next two chapters.

CHAPTER 3

DEVELOPMENT OF STATISTICALLY AND ENERGY EFFICIENT ALT PLANS

This chapter presents a new methodology for designing statistically and energy efficient ALT plans. Section 3.1 briefly introduces the traditional method for optimum design of ALT plans and identifies the limitation of the traditional method with respect to energy consumption. Section 3.2 addresses two methods to evaluate the energy consumption of ALT experiment. These methods reveal the relationship among adopted ALT plan, associated control strategy and energy consumption. In Section 3.3, a conceptual framework for designing statistically and energy efficient ALT plans is developed. Based on this framework, a sequential ALT design approach for planning such ALT experiments is formulated mathematically. The computational algorithms and simulation tools developed in this research to handle such complex experimental design problems are provided in Section 3.4.

3.1 Traditional Method for Optimum Design of ALT Plans

Most research on the design of optimum ALT plans is focused only on determining the levels of stress loadings and unit allocation to each stress level to minimize the uncertainty in reliability estimates under some constraints. Let $G(\underline{\Theta}, Z_0)$ be the interested product reliability index under normal operating conditions Z_0, where $\underline{\Theta}$ is the parameter vector of the ALT model. Considering singly Type I censoring with censoring time $\tau^{(0)}$, the statistical precision of estimated reliability index $\hat{G}(\hat{\underline{\Theta}}, Z_0 \mid [\Omega^{(0)}, \tau^{(0)}])$ depends on the adopted ALT plan $\Omega^{(0)} = [Z_1^{(0)}, ..., Z_k^{(0)}, n_1^{(0)}, ..., n_k^{(0)}]$, where $n_i^{(0)}$ is the number of test units allocated to stress loading $Z_i^{(0)}$, and $\hat{\underline{\Theta}}$ is the maximum likelihood (ML) estimate of $\underline{\Theta}$. According to ML theory, $\hat{\underline{\Theta}}$ asymptotically follows the multivariate normal distribution $MVN(\hat{\underline{\Theta}}, \hat{\Sigma}_{\hat{\Theta}})$ with the variance-covariance matrix $\hat{\Sigma}_{\hat{\Theta}} = \left[E\left[-\frac{\partial^2 L(\Theta \mid \text{data})}{\partial \Theta \partial \Theta^T} \right] \right]^{-1}$ evaluated at $\hat{\underline{\Theta}}$, in which $E[\cdot]$ denotes expectation and $L(\Theta \mid \text{data})$ is the log-likelihood function of ALT data (Nelson, 1990).

Mathematically, the optimum experimental design problem $(P1)$ can be formulated as:

$$\Omega^{(0)*} = \arg\min_{\Omega^{(0)}} Asvar\left(\hat{G}(\hat{\underline{\Theta}}, Z_0 \mid [\Omega^{(0)}, \tau^{(0)}]) \right)$$

$$P1: \quad \text{Subject to} \quad H_1(\Omega^{(0)}, \tau^{(0)}) \leq 0, \tag{3.1}$$
$$H_2(\Omega^{(0)}, \tau^{(0)}) = 0,$$
$$n_1, ..., n_k \in \{1, 2, ...\},$$

where the asymptotic variance:

$$Asvar\left(\hat{G}(\hat{\underline{\Theta}}, Z_0 \mid [\Omega^{(0)}, \tau^{(0)}]) \right) = \left[\frac{\partial \hat{G}}{\partial \Theta} \right]^T \hat{\Sigma}_{\hat{\Theta}} \left[\frac{\partial \hat{G}}{\partial \Theta} \right], \tag{3.2}$$

quantifying the statistical efficiency of the ALT plan is to be minimized; functions $H_1(\Omega^{(0)}, \tau^{(0)})$ and $H_2(\Omega^{(0)}, \tau^{(0)})$ consist of all inequality and equality constraints, respectively, such as the ranges of stress amplitudes, the maximum ramp rate, and possible types of stress loadings (e.g., see Figure 1-1). Note that $\tau^{(0)}$ is not a decision variable. Usually, such optimum ALT plans may not perform well in terms of energy use. To the best of our knowledge, this gap has already become a barrier for technology innovation, but a viable solution has not been reported yet.

3.2 Energy Consumption of an ALT Experiment

To quantify the total energy consumption of an ALT experiment, let $\mathbb{P}(t;Z_i)$ be the instantaneous power consumption of the test equipment associated with stress loading Z_i, $i=1,2,3,\ldots,m$, at time t, and $\mathbb{EN}_i(t;Z_i)$ be the energy consumption attributed to Z_i up to time t. If $\mathbb{P}(t;Z_i)$ can be measured over time (e.g., with a power meter), the total energy consumption of the ALT experiment with duration of τ can be obtained by integrating instantaneous power consumption for each of the stress loadings up to τ :

$$\sum_{i=1}^{m} \mathbb{EN}_i(\tau;Z_i) = \sum_{i=1}^{m} \int_{0}^{\tau} \mathbb{P}(t;Z_i)\, dt. \tag{3.3}$$

For example, the values of daily energy consumption of the three cyclic-stress loadings shown in Figure 1-2 are 18.4 kW·h, 26.3 kW·h, and 29.1 kW·h, respectively, and the total energy consumed during this short time period is 73.8 kW·h.

Moreover, different control strategies (e.g., on/off or proportional-integral-derivative (PID) controllers) adopted by the test equipment may result in considerable differences in energy consumption. For instance, an on/off controller normally consumes more energy than a PID controller due to frequent and abrupt switching between on and off states. Besides, even for a specific type of controller, the setting of control parameters has significant impact on accurate profile generation and energy consumption. For example, consider an environmental chamber controlled by a PID controller with three control parameters K_p, K_i and K_d. When the proportional parameter K_p is large, the controller tends to be more sensitive to the tracking error between the required temperature profile and the chamber's actual response. In this case, the controller may cause high-frequency alternations between the heating and cooling modes, which increase the power consumption and tracking error during test. On the other hand, if K_p is too small, slower control actions may reduce the energy consumption to some extent but meanwhile may cause a significant increase in tracking error. Unlike K_p, the integral parameter K_i responds to the accumulated error and large K_i may reduce the steady-state tracking error. However, if its value is too large, the system response may exhibit strong oscillations, which can significantly increase the energy consumption as well. Moreover, the differential term K_d can help stabilize the closed-loop system, but an excessively large K_d may cause sluggish system response, thus increase the tracking error.

To demonstrate the impact of control parameters on the energy consumption of an ALT experiment, Table 3-1 presents different PID designs for a *two-level-modified-step-stress* loading where test equipment's transfer function is assumed to be $H(S) = 38.8/(2475s+1)$ (Yang et al., 2006) and its instantaneous power consumption can be expressed as: $\mathbb{P}(u(t;Z_i)) = 1 + u^2(t;Z_i)$. The tracking errors in terms of root mean squared error (RMSE) for actual stress loadings generated by these controllers are also presented. One can see that, for the given stress loading profile, by solely adjusting the values of control parameters, the energy use changes from 5,830,422 to 36,681,426 units of energy. In the worst case, when $K_p = 100$, $K_i = 2$, $K_d = 10$, the test equipment can consume 5 times as much energy as the benchmark PID controller. The tracking errors vary from 0.0668 to 2.5681 degree Celsius, which may potentially damage the test units (e.g., overheat) and affect the statistical estimation precision of the experiment due to the difference between the desired stress loading and the one the test units are actually exposed to.

Table 3-1: Influence of PID controller design on energy consumption of a test

Stress loading	$[K_p, K_i, K_d]$	Energy	RMSE
$25°C$ — Ramp rate $=2°C/min$ — $75°C$ — Z_1^* $150°C$ — Stop test; 0 … 144.7hrs τ Time; $\tau = 199.4$ hours.	[4.9904, 0.75, 0.01.] (Benchmark PID controller)	5,968,637 units of energy	0.1529 $°C$
	[20, 0.5, 10]	6,883,961 units of energy (Increase 15.34%)	0.0668 $°C$
	[100, 2, 10]	36,681,426 units of energy (Increase 515%)	2.5681 $°C$
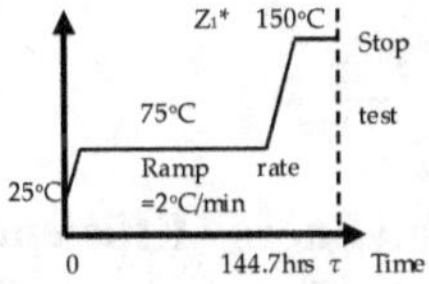	[20, 2, 10]	6,949,120 units of energy (Increase 16.43%)	0.0796 $°C$
	[0.01, 0.75, 0.01]	7,355,858 units of energy (Increase 23.24%)	0.8694 $°C$
	[0.01, 0.01, 0.01]	5,830,422 units of energy (Decrease 2.32%)	0.9378 $°C$
	[0.01, 0.01, 100]	6,148,244 units of energy (Increase 3.01%)	1.3304 $°C$

If it is flexible to choose control strategies, the controller design based on the model of the chamber (e.g., transfer function) and the stress loadings could effectively reduce the energy use and tracking error of ALT experiment, which will introduce another important set of decision variables. After all, different control strategies have different capabilities that impose certain constraints on the stress loadings to be implemented.

Let $u(t;Z_i)$ be the input (i.e., control signal) to the test equipment at time t. Then, the total energy consumption up to time τ can be expressed more generally as:

$$\sum_{i=1}^{m} \mathbb{EN}_i(\tau, \{u(t;Z_i), 0 \leq t \leq \tau\}) = \sum_{i=1}^{m} \int_0^{\tau} \mathbb{P}(u(t;Z_i)) \, dt, \tag{3.4}$$

where $\mathbb{P}(u(t;Z_i))$ is the instantaneous power consumption depending on the control signal $u(t;Z_i)$. Clearly, the energy consumption has a functional relationship with the parameters (e.g., high/low temperatures and ramp rate) of applied stress loadings Z_i, number of stress levels m, and the test duration τ, which are among the most important decision variables in planning ALT experiments. By manipulating these decision variables, it is possible for thousands of product developers to reduce the energy consumption in product developments using ALT.

3.3 Methodology for Planning Precise and Energy Efficient ALT

The new framework that interfaces the optimum design of ALT plans and active energy reduction is shown in Figure 3-1. Prior to ALT planning, a pilot ALT experiment will be conducted to identify the underlying ALT model and estimate the baseline model parameters. The double-loop optimization architecture can be utilized when it is flexible to determine a controller for the test equipment and/or perform optimization in experimental design. For this new experimental design methodology, the estimation precision of each candidate ALT plan is evaluated and the corresponding controller capable of providing the required stress loadings is designed and evaluated to find the total energy consumption of the resulting experiment. There are two ways to planning such ALT experiments: (1) optimization of statistical efficiency followed by energy minimization, and (2) simultaneous optimization of statistical efficiency and energy consumption. The simultaneous optimization method belongs to a large-scale optimization problem. To demonstrate the new experimental design methodology, Section 3.3.1 is focused on the first method, which is relatively easier to implement in practice. The simultaneous optimization method will be investigated in Chapter 4.

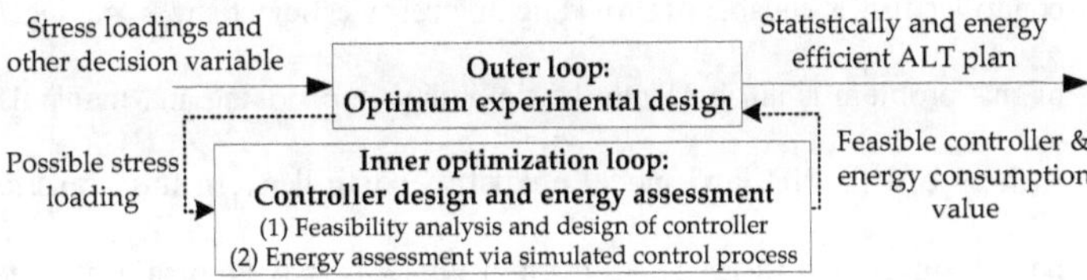

Figure 3-1: Proposed methodology for planning statistically and energy efficient ALT

3.3.1 Sequential Optimization Approach for Designing Energy and Statistically Efficient ALT Plans

For ease of presentation, we provide the following definitions (Liao and Elayed, 2010). Let $\mathbb{X}$ and $\mathbb{Y}$ be the variance-covariance matrices of $\hat{\underline{\Theta}}$ obtained by two ALT plans, respectively. The two ALT plans are statistically equivalent, in two extreme cases, if:

(1) $\mathbb{X} = \mathbb{Y}$ (Type III statistically equivalent ALT plan), or

(2) $\left[\frac{\partial \hat{G}}{\partial \hat{\underline{\Theta}}}\right]^{T} \mathbb{X} \left[\frac{\partial \hat{G}}{\partial \hat{\underline{\Theta}}}\right] = \left[\frac{\partial \hat{G}}{\partial \hat{\underline{\Theta}}}\right]^{T} \mathbb{Y} \left[\frac{\partial \hat{G}}{\partial \hat{\underline{\Theta}}}\right]$ (General statistically equivalent ALT plan).

Compared to Type III statistically equivalent ALT plans, general statistically equivalent ALT plans are easier to find, thus are implemented in the sequential optimization procedure proposed in study.

In a particular scenario where the test equipment are treated as "black boxes", the physical model of the test equipment, such as the transfer function, can be obtained via a system identification tool often used in controls area. Meanwhile, the optimal design of ALT plan will be performed sequentially that determines an optimum ALT plan in terms of estimation precision followed by energy minimization through searching for a feasible controller that is capable of providing an energy efficient stress loading. The complexity of this problem is attributed to the flexibility in choosing and manipulating controllers such as on/off, PID, and model predictive controllers. In this scenario, instantaneous power consumption $\mathbb{P}(u(t;Z_i))$ given in Eq. (3.4) may be obtained based on the control signal $u(t;Z_i)$ instead of using an empirical approach. In other words, $\mathbb{P}(u(t;Z_i))$ depends on the control signal $u(t;Z_i)$ generated by the controller designed. For example, $\mathbb{P}(u(t;Z_i))$ could be a quadratic function of $u(t;Z_i)$: $\mathbb{P}(u(t;Z_i)) = c_0 + c_1 u^2(t;Z_i)$, where c_0 is the power consumption of the equipment while idle and c_1 is a constant.

Let $\Omega^{(1)}$ be $[Z_1,...,Z_m,n_1,...,n_m]$ containing all the stress loadings and unit allocations in the desired statistically and energy efficient ALT plan. Note that τ becomes a decision variable. Mathematically, we will solve the following experimental design problem:

Step One: Solve problem P1 (see Eq. (3.1)):

$$\Omega^{(0)*} = [Z_1^{(0)*},...,Z_k^{(0)*},n_1^{(0)*},...,n_k^{(0)*}]$$

$$\rightarrow$$

(Statistically efficient benchmark ALT plan)

Step Two: Solve problem P2:

$$\min_{[\Omega^{(1)},\tau]\&\text{controller}} \sum_{i=1}^{m} \mathbb{EN}_i(\tau,\{u(t;Z_i),0 \le t \le \tau\}) = \sum_{i=1}^{m} \int_0^{\tau} \mathbb{P}(u(t;Z_i))\, dt,$$

$$\text{Subject to} \quad Asvar\left(\hat{G}(\hat{\Theta},Z_0\,|\,[\Omega^{(1)},\tau])\right) = Asvar\left(\hat{G}(\hat{\Theta},Z_0\,|\,[\Omega^{(0)*},\tau^{(0)}])\right), \quad (3.5)$$

and other constrains including those in Problem *P1*.

$\rightarrow$ <u>Statistical and energy efficient ALT plan</u>: $[\Omega^{(1)*},\tau^*]$ and the best setting of the controller.

In this sequential optimization problem, after obtaining a statistically efficient ALT plan $[\Omega^{(0)*},\tau^{(0)}]$ as shown in Eq. (3.1), algorithms are needed in seeking a general equivalent ALT plan $[\Omega^{(1)*},\tau^*]$ while choosing and manipulating a feasible controller that minimizes the energy use of the equivalent ALT plan.

3.4 Computational Algorithms for Sequential Optimum ALT Design

Decision making on desired ALT plans belongs to a large-scale optimization problem. Especially, statistical efficiency is partially attributed to the applied stress loadings which are tightly interrelated with the total energy consumption that is associated with the control strategy designed for the test equipment. Their complex interrelationship needs to be explored extensively before being mathematically formulated. In addition, advanced energy assessment tools and optimization algorithms are needed in effectively manipulating the associated decision variables. Since PID control is one of the most widely used strategies in instrumentation and control, algorithms based on PID controllers are developed in this section. Because PID controllers are either adjustable or unadjustable, algorithmic developments for both of these two cases will be addressed.

3.4.1 Case 1: Algorithm for Test Equipment with an Adjustable PID Controller

Figure 3-2 shows the structure of a piece of test equipment with a conventional PID controller. The control signal is: $u(t) = K_p\, e(t) + K_i \int_0^t e(v)\, dv + K_d\, [de(t)/dt]$, where $e(t) = Z_i(t) - \tilde{Z}_i(t)$ is the error between the desired stress loading Z_i and the actual output $\tilde{Z}_i$ of the test equipment. For the case where the parameters, K_p, K_i and K_d, are adjustable, the desired ALT plan under the corresponding constraints according to Eq. (3.5) becomes:

$$[\Omega^{(1)*}, \tau^*, K_p^*, K_i^*, K_d^*] = \operatorname*{arg\,min}_{[\Omega^{(1)}, \tau], K_p, K_i, K_d} \sum_{i=1}^{m} \mathbb{EN}_i(\tau, \{u(t; Z_i), 0 \le t \le \tau\}). \qquad (3.6)$$

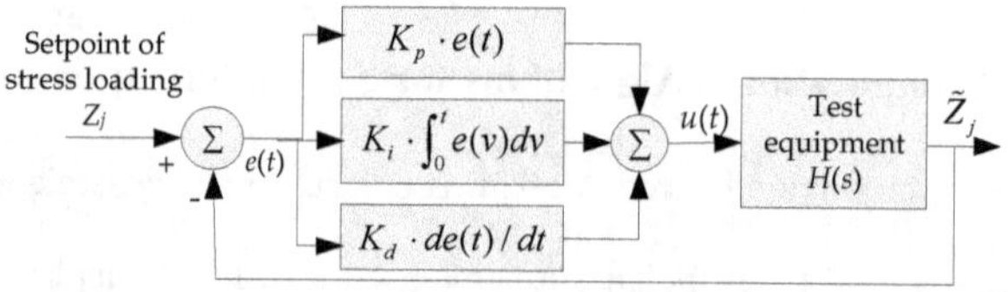

Setpoint of stress loading Z_j → Σ (+ / –), $e(t)$ → $K_p \cdot e(t)$, $K_i \cdot \int_0^t e(v)dv$, $K_d \cdot de(t)/dt$ → Σ → $u(t)$ → Test equipment $H(s)$ → $\tilde{Z}_j$

Figure 3-2: A controlled system with a conventional PID controller

To solve this problem, a set of design tools is developed and coded in MATLAB. For each candidate ALT plan providing the desired reliability estimation precision, we initialize the simulation system using PID parameters $K_p = K_{p0}$, $K_i = K_{i0}$, $K_d = K_{d0}$ obtained via the Ziegler-Nichols Method (Åström and Hägglund, 1995)), and use a

Bisection Method (see Figure 3-3) to determine the best values of K_p, K_i and K_d capable of tracking the stress profile specified by the ALT plan in terms of RMSE. If feasible, the model for the controlled system with the resulting PID controller is simulated to evaluate the energy consumption of this candidate ALT plan. A noise generator is included in the simulation program (see Figure 3-4) to simulate the fluctuation of the surrounding environment (e.g., ambient temperature, relative humidity, power supply) of the test equipment. The simulation program also provides a graphical user interface to setup the stress loading profile specified by ALT plan (see Figure 3-5). By comparing the energy consumption of all candidate plans, the ALT plan consuming the least amount of energy while providing the most precise reliability estimate will be selected. In addition to the desired ALT plan, other important outputs: (1) system response vs. reference signal, (2) control signal, (3) energy consumption, and (4) the tracking error between the actual system response and the desired stress profile, are also generated (see Figure 3-6).

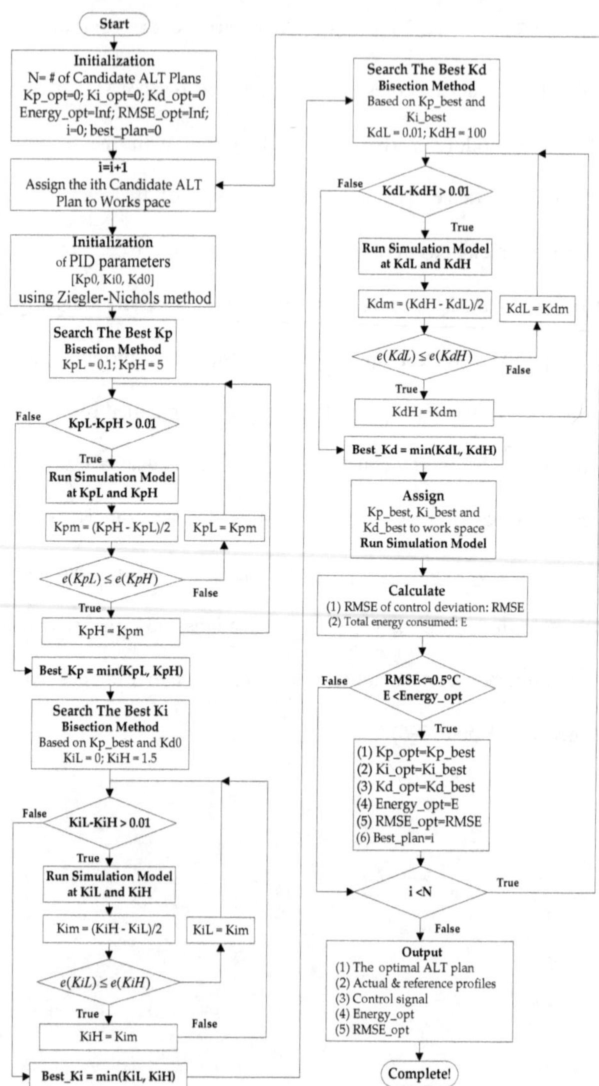

Figure 3-3: Flow diagram of the Bisection search algorithm for tuning a PID controller

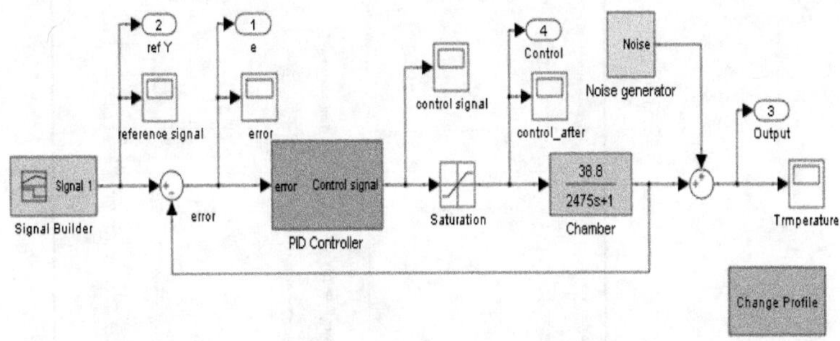

Figure 3-4: Simulink model for the controlled test equipment

Figure 3-5: User interface for stress loading profile setup

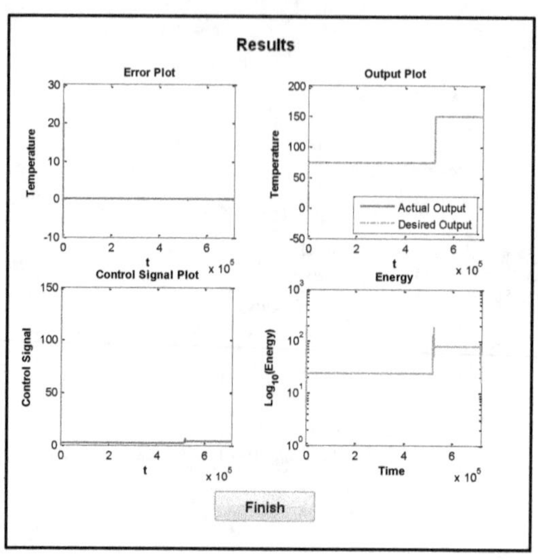

Figure 3-6: Outputs of the Simulink model

3.4.2 Case 2: Algorithm for Test Equipment with an Unadjustable PID Controller

In practice, the parameters of PID controller of a piece of test equipment may be fixed by the manufacturer. In such a case, given different stress loadings, the test equipment may exhibit different features in response time and tracking errors, etc., which are quite important to actual stress exposure of test units in ALT. In order to ensure the test equipment's performance in providing the desired stress-loading profile, the input-weighting method (Eitelberg, 1987) can be implemented to regulate the equipment's response. Figure 3-7 shows the structure of a system with an input-weighting PID controller, where three input-weighting factors F_p, F_i and F_d are included in the control loop without changing the given PID controller.

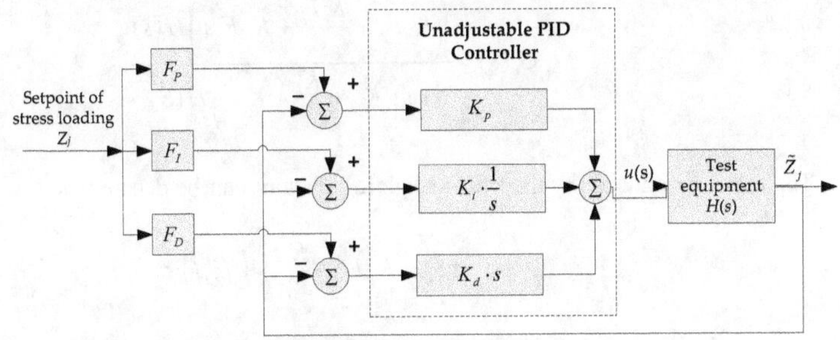

Figure 3-7: A control system with an input-weighting PID controller

Let $H(s)$ be the transfer function of the test equipment. The transfer function of the given PID controller of the test equipment can be expressed as:

$$G_{PID}(s) = K_p + \frac{K_i}{s} + K_d s = \frac{K_d s^2 + K_p s + K_i}{s}. \tag{3.7}$$

Then, the transfer function of the closed-loop system with the given controller is:

$$G_c(s) = \frac{\tilde{Z}_j(s)}{Z_j(s)} = \frac{G_{PID}(s)H(s)}{1 + G_{PID}(s)H(s)}, \tag{3.8}$$

where $Z_j(s)$ is the desired stress-loading profile specified by certain ALT plan and $\tilde{Z}_j(s)$ are the actual stress-loading profile created by test equipment. Eq. (3.8) indicates that the design of PID controller plays a decisive role in determining the performance of the closed-loop system. As shown in Figure 3-7, after introducing three input-weighting factors F_p, F_i and F_d, the transfer function of the modified closed-loop system can be expressed as:

$$G_c'(s) = \frac{\left(K_p F_p + \dfrac{K_i F_i}{s} + K_d F_d s\right) H(s)}{1 + \left(K_p + \dfrac{K_i}{s} + K_d s\right) H(S)} .$$ (3.9)

The terms introduced by the input-weighting factors can be denoted as:

$$G_{PID}'(S) = K_p F_p + \frac{K_i F_i}{s} + K_d F_d s .$$ (3.10)

As a result, Eq. (3.9) can be rewritten as (Mantz and Tacconi, 1990):

$$G_c'(s) = \frac{G_{PID}'(s) H(s)}{1 + G_{PID}(s) H(s)} = \frac{G_{PID}'(s)}{G_{PID}(s)} G_c(s) .$$ (3.11)

In this case, because the values of K_p, K_i and K_d are fixed, the given PID controller may not provide the desired control actions for achieving a specific stress profile. Based on Eq. (3.11), by properly choosing the values of F_p, F_i and F_d, the undesirable effects caused by the dominant poles in the original closed-loop test system (brought by the test equipment characteristics and the fixed PID controller) can be cancelled by the zeros of $G_{PID}'(s)$ (Eitelberg, 1987; Mantz and Tacconi, 1990). Therefore, the performance of test equipment can be improved with respect to both energy consumption and control performance. However, to minimize the steady-state error between the desired stress loading and the actual output of test equipment, the value of integral weighing parameter F_i needs to be fixed as value of one, and the performance of test equipment can be improved by only adjusting the values of F_p and F_d.

Another practical problem is that since the PID control parameters are fixed in this case, given a stress-loading profile, the resulting closed-loop system could be either under-damped or over-damped. When the original closed-loop system is over-damped, the given PID controller causes sluggishness in the test equipment response, which may result in large tracking error. By adjusting the values of F_p and F_d, system response speed can be accelerated and therefore reduce the tracking error to create a more precise test environment. In situation that the original closed-loop system is under-damped, large amplitude overshoot and high frequency oscillations could be caused by the given PID controller. The overshoot amplitude and frequency of oscillations exist in system response can also be effectively reduced through manipulating the values of the two input-weighting factors. A simulation tool (see Figure 3-8) and a computational procedure based on the active-set algorithm are developed to search for the best input-weighting factors that optimize the performance of the test equipment with respect to tracking accuracy and finding the desired ALT plan:

$$[\Omega^{(1)*}, \tau^*, F_p^*, F_d^*] = \underset{[\Omega^{(1)}, \tau], F_p, F_d}{\arg\min} \sum_{i=1}^{m} \mathbb{EN}_i(\tau, \{u(t; Z_i), 0 \leq t \leq \tau\}), \tag{3.12}$$

under the corresponding constraints.

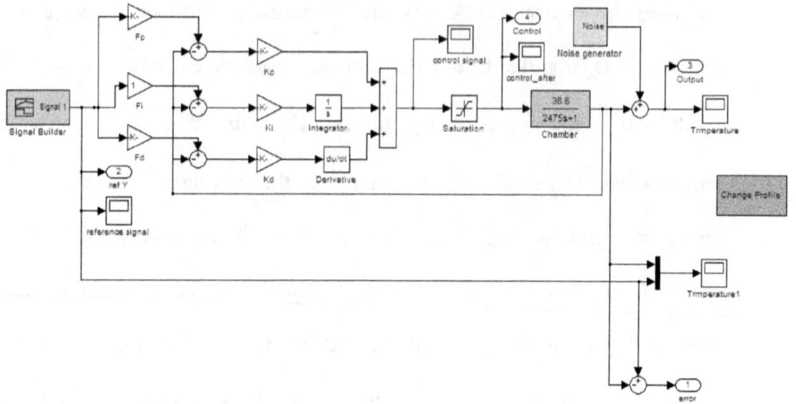

Figure 3-8: Simulink model for test equipment with an input-weighting PID controller

CHAPTER 4

SIMULTANEOUS OPTIMIZATION METHOD FOR DESIGNING STATISTICALLY AND ENERGY EFFICIENT ALT PLANS

For the sequential ALT planning procedure presented in Chapter 3, the choice of the stress loading in the benchmark ALT plan would put certain restrictions on the resulting ALT plan. As a result, the resulting ALT plan could be locally optimal. It is desired to develop a fully integrated approach and computational tool for designing ALT plans that are globally optimal in terms of statistical estimation precision and energy efficiency. To solve this problem, a simultaneous optimization framework is developed in this dissertation. This chapter presents the mathematical formulations of the proposed simultaneous optimization approach along with algorithms for implementing this approach.

4.1 Mathematical Formulation of Simultaneous Optimization Method for ALT Design

As an alternative and extension to the sequential optimization approach introduced in Section 3.3.1, the new optimization framework aims to simultaneously minimize the uncertainty in reliability estimates and energy consumption. As discussed in Chapter 3, the reliability estimation precision can be affected by the type and level of adopted stress-loading as well as the test duration. The energy consumption and tracking performance have a complex relationship with both stress-loading profile and the adopted control strategy. Since tracking error could potentially affect the statistical estimation precision of an ALT experiment due to the difference between the desired stress loading and the one the test units are actually exposed to, the tracking performance of test equipment should also be addressed and improved in the ALT design stage. Hence, we consider three objectives in the proposed simultaneous optimization framework:

1) Minimize the total energy consumption of ALT experiment

2) Minimize the uncertainty in reliability estimates

3) Minimize the tracking error between the desired and the actual stress conditions

The total energy consumption during test can be estimated using Eq. (3.4), where we assume the instantaneous power consumption $\mathbb{P}(u(t;Z_i))$ is a function of control signal $u(t;Z_i)$ corresponding to stress loading profile Z_i. Eq. (3.2) can be used to quantify the uncertainty in reliability estimation. Moreover, the tracking performance of designed ALT plans can be evaluated using $RMSE = \sqrt{\dfrac{\sum_{j=1}^{N} e(t_j)^2}{N}}$, where $e(t_j) = Z_i(t_j) - \tilde{Z}_i(t_j)$ ($i = 1,...,m$) represents the error between the desired stress loading Z_i and the actual output $\tilde{Z}_i$ created by test equipment and $0 \le t_j \le \tau_m$ is the j th sample point.

Since all the three objectives are closely related to the adopted stress loading profile and control strategy for the test equipment, the decision variables of this multi-objective optimization problem should include:

- Z_i : stress level i in an ALT experiment, $i = 1,...,m$

- n_i : number of test units allocated at Z_i, $i = 1,...,m$

- τ_i : stress level change time at Z_i, $i = 1,...,m-1$

- τ : censoring time of an ALT experiment and $\tau_m = \tau$

- $\underline{C}$: a set of control parameters adopted by the test equipment. For example $\underline{C}$ for

  a test equipment with a PID controller would be $[K_p, K_i, K_d]$

Let $\underline{X}$ be $[Z_1,...,Z_m, n_1,...,n_m, \tau_1,...\tau_{m-1}, \tau, \underline{C}]$ containing all the decision variables in

the desired statistically and energy efficient ALT plan. Mathematically, the interested

optimization problem with three objectives can be formulated as:

$$\underline{X}^* = \arg\min_{\underline{X}} \{f_1(\underline{X}), f_2(\underline{X}), f_3(\underline{X})\}$$

$$\text{Subject to} \quad H_1(\underline{X}) \leq 0,$$
$$H_2(\underline{X}) = 0, \tag{4.1}$$
$$n_1,...,n_k \in \{1, 2,...\},$$

where

$$f_1(\underline{X}) = \sum_{i=1}^{m} \mathbb{EN}_i(\tau_i, \{u(t; Z_i), 0 \leq t \leq \tau_i\}) = \sum_{i=1}^{m} \int_0^{\tau_i} \mathbb{P}(u(t; Z_i)) \, dt, \tag{4.2}$$

$$f_2(\underline{X}) = Asvar\left(\hat{G}(\hat{\Theta}, Z_0 \mid [\underline{X}])\right), \tag{4.3}$$

$$f_3(\underline{X}) = RMSE = \sqrt{\frac{\sum_{j=1}^{N} \left[Z_i(t_j) - \tilde{Z}_i(t_j)\right]^2}{N}}, 0 \leq t_j \leq \tau, \tag{4.4}$$

and functions $H_1(\underline{X})$ and $H_2(\underline{X})$ represent all the inequality and equality constraints that

must be satisfied in the experimental design process.

In this problem, the three objectives are conflicting to each other to some extent. For instance, in order to increase the accuracy and precision of reliability estimation, a commonly utilized strategy is to extend the test duration. However, longer test time can result in more energy consumption of the experiment. On the other hand, when we attempt to reduce energy use by adjusting the PID control parameters, slower control actions are always preferred over faster ones. But the slow control actions may enlarge tracking errors. In this case, the accuracy and precision of reliability estimation may also be jeopardized by the inconsistency between the desired and actual test conditions. Unlike a single-objective optimization problem, a unique optimal solution which optimizes all objectives may not exist in such a multi-objective optimization problem. As a result, trade-offs among all the objectives are needed for solving this simultaneous optimization problem.

4.2 Computational Algorithms

There are a variety of approaches for solving multi-objective optimization problems. One classical solution is to convert those objectives to a composite optimization criterion so that this problem can be solved as a single-objective optimization problem. However, the obtained optimal solution would significantly rely on the defined composite optimization criterion which reflects the test designer's preference. Without a correct and comprehensive understanding about the characteristics of original objective functions and their interactions, using a composite optimization criterion may be questionable, thus results in an undesired solution. Another widely used solution is to use Multi-objective Evolutionary Algorithms (MOEAs) to deal with such multi-objective optimization problems. MOEAs require no information about the preference of decision makers and can provide a set of Pareto-optimal solutions in one single simulation run (Deb et al., 2002; Liao and Li, 2008). For the proposed problem (see Eq. (4.1)), the three objective functions have a complex relationship, so it is desired to obtain more solutions with different trade-offs among those optimization criteria. Since 1980s, a large amount of research has been conducted regarding the MOEAs. Many algorithms have been developed for solving multi-objective optimization problems, such as Vector Evaluated GA (Schaffer, 1985), Non-dominated Sorting Genetic Algorithm (NSGA) (Srinivas and Deb, 1995), and NSGA-II (Deb et al., 2002). In this chapter, the simultaneous optimization problem for the design of statistically and energy efficient ALT plans is solved using a controlled elitist non-dominated sorting genetic algorithm (Deb, 2001; Deb and Goel, 2001). This algorithm is an improved version of NSGA-II and can effectively preserve both elitism and diversity among possible solutions.

4.2.1 Controlled Elitist Non-dominated Sorting Genetic Algorithm (Deb, 2001; Deb and Goel, 2001)

For better understanding of the controlled elitist non-dominated genetic algorithm, we need first understand the concept of domination. Assume $\underline{x}_1$ and $\underline{x}_2$ are two solutions of a multi-objective optimization problem with m objectives: $f_i(\underline{x})$, $i \in \{1, 2, ..., m\}$. Solution $\underline{x}_1$ is said to dominate $\underline{x}_2$, if it satisfies the following two conditions simultaneously:

1. $f_i(\underline{x}_1)$ is not worse than $f_i(\underline{x}_2)$ for all $i \in \{1, 2, ..., m\}$;

2. $f_i(\underline{x}_1)$ is better than $f_i(\underline{x}_2)$ for at least one $i \in \{1, 2, ..., m\}$.

If we cannot find any other solution that dominates a solution $\underline{x}$, then $\underline{x}$ is referred to as a non-dominated solution. According to this concept, we can compare and sort all solutions generated in one generation and then identify elites from the current generation based on the ranks of the solutions.

In the controlled elitist non-dominated sorting genetic algorithm, a random or pre-specified initial parent population P_0 is first created. All solutions contained in P_0 are sorted using the fast non-dominated sorting approach based on the corresponding values of objective functions. Each solution is then assigned a rank which is equal to its non-dominated level. Low-rank solutions are preferred over the ones with higher ranks. In order to generate the next generation, a child population Q_0 is produced using selection, crossover, and mutation operators. Combining P_0 and Q_0, a combined population $R_0 = P_0 \cup Q_0$ is obtained. Next, R_0 is sorted and classified into different non-dominated fronts F_k ($k = \{1, 2, ...\}$) using the fast non-dominated sorting approach, where F_1 is the best non-dominated set consisting of the best solutions in the combined population R_0, F_2 is the second best non-dominated set, and so forth. The new parent population P_1 will be generated by selecting solutions from all non-dominated fronts F_k according to a controlled approach (Deb and Goel, 2001).

In the controlled approach, the maximum number of solutions in the k th front F_k allowed to be added into the new parent population P_1 is:

$$n_k = N \frac{1-r}{1-r^K} r^{i-1},$$
(4.5)

where N is the size of P_1 and $r \in (0,1)$ is the reduction rate. Since r is smaller than one, the maximum allowable number of solutions to be selected from each front is decreasing as the front index increases. Let n_{Fk} represents the number of solutions contained in the k th non-dominated front. If n_{Fk} is greater than n_k, only n_k solutions will be chosen from F_k using the crowded-comparison operator, which prefers the solutions with the lower rank or the solutions located in the less crowed regions. On the other hand, if n_{Fk} is smaller than n_k, all n_{Fk} solutions will be included in the new parent population and the remaining $n_k - n_{Fk}$ slots will be filled by the solutions belonging to the next front F_{k+1}. In this case, $n_{k+1} = n_{k+1} + (n_k - n_{Fk})$. This process is continued until all N solutions in P_1 are selected, and then P_1 will repeat the experience of P_0 to generate the next generation. This algorithm will stop when one of stopping criteria is met. Normally, the stopping criteria include running time limit, maximum number of generations, and so on. The flow diagram of this algorithm is provided in Figure 4-1.

As stated in Section 3.4, PID control is one of the most widely used strategies adopted by reliability testing equipment, simultaneous optimization algorithms based on PID controllers are developed to design statistically and energy efficient ALT plans. In the next two sections, algorithmic developments for cases with adjustable and unadjustable PID controllers will be investigated, respectively.

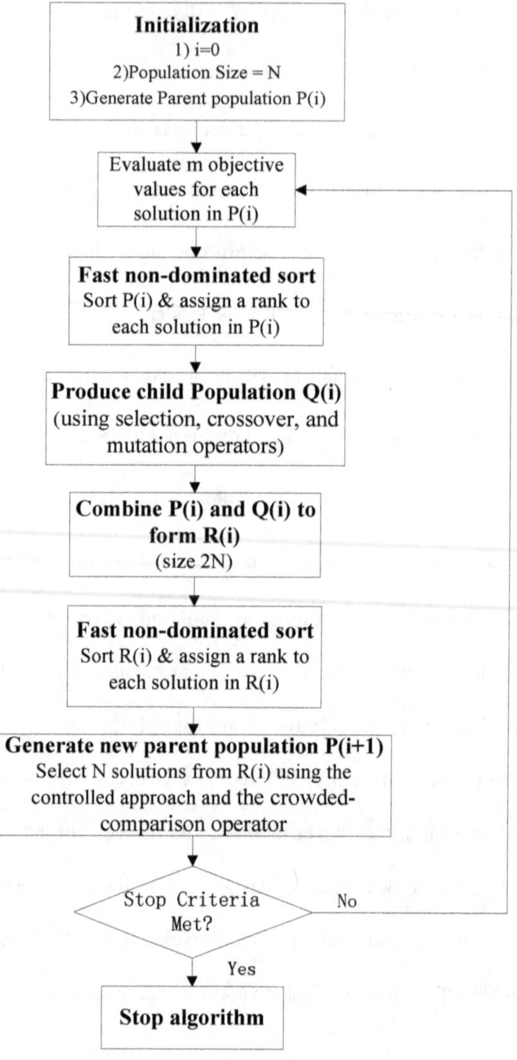

Figure 4-1: Flow diagram of the controlled elitist non-dominated sorting genetic algorithm

4.2.2 Case 1: Algorithm for Test Equipment with an Adjustable PID Controller

When the test equipment has an adjustable PID controller (see Figure 3-2), the user can directly regulate the operation condition of test equipment by specifying the values of three PID control parameters, K_p, K_i and K_d. Therefore, in this situation, decision variables included in $\underline{C}$ of Eq. (4.1) representing the control parameters adopted by the test equipment should be $[K_p, K_i, K_d]$. In this case, the simultaneous optimization problem becomes:

$$\underline{X}^* = \arg\min_{\underline{X}}\{f_1(\underline{X}), f_2(\underline{X}), f_3(\underline{X})\}$$

$$\text{Subject to} \quad \begin{aligned} &H_1(\underline{X}) \leq 0, \\ &H_2(\underline{X}) = 0, \\ &K_p, K_i, K_d \geq 0 \\ &n_1, ..., n_k \in \{1, 2, ...\}, \end{aligned} \qquad (4.6)$$

where $\underline{X} = [Z_1, ..., Z_m, n_1, ..., n_m, \tau_1, ..., \tau_{m-1}, \tau, K_p, K_i, K_d]$. To solve this problem, a set of computational algorithms involving the simulation model for the controlled test equipment are developed using MATLAB in evaluating the three objective functions for each possible solution. This objective function evaluation algorithm is then embedded into the controlled elitist non-dominated genetic algorithm to obtain the statistically and energy efficient ALT plan.

4.2.3 Case 2: Algorithm for Test Equipment with an Unadjustable PID Controller

As discussed in Section 3.4.2, when manufacturers choose to fix the parameters of PID controller, we will use input-weighting method to regulate and improve the equipment's response according to the given stress-loading profiles without changing the built-in PID controller (see Figure 3-7). In this case, decision variables included in $\underline{C}$ of Eq. (4-1) should be $[F_p, F_i, F_d]$. Since the value of integral weighing parameter F_i should be fixed at 1 to minimize the steady-state error between the desired stress loading and the actual output of test equipment, we will not treat F_i as a decision variable and thus $\underline{C} = [F_p, F_d]$. Under this scenario, the simultaneous optimization method for designing statistically and energy efficient ALTs can be expressed as:

$$\underline{X}^* = \arg\min_{\underline{X}}\{f_1(\underline{X}), f_2(\underline{X}), f_3(\underline{X})\}$$
$$\text{Subject to}\quad H_1(\underline{X}) \leq 0,$$
$$H_2(\underline{X}) = 0, \tag{4.7}$$
$$F_p, F_d \geq 0$$
$$n_1, \ldots, n_k \in \{1, 2, \ldots\},$$

where $\underline{X} = [Z_1, \ldots, Z_m, n_1, \ldots, n_m, \tau_1, \ldots, \tau_{m-1}, \tau, F_p, F_d]$. Based on the controlled elitist non-dominated genetic algorithm, a computational procedure involving simulation is developed for test system with an unadjustable PID controller using Matlab.

CHAPTER 5

RESULTS AND SENSITIVITY ANALYSIS

In this chapter, we illustrate the feasibility and potential of designing statistically and energy efficient ALT plans. In Section 5.1, we design statistically and energy efficient ALT experiment for exponential lifetime distributions using the proposed sequential ALT design approach. In Section 5.2, the simultaneous optimization procedure is implemented to design optimum ALT plan with respect to both energy consumption and statistical efficiency for estimating a specified quantile of the Weibull distribution at use condition. Section 5.3 compares the performance of the proposed sequential and simultaneous optimum ALT design procedures regarding both energy consumption and statistical estimation precision. The robustness of optimum ALT plans designed using sequential ALT planning procedure is studied in Section 5.4.

5.1 Application of Sequential Optimization ALT Design Method

Considering an exponential ALT model with a log-linear life-stress relationship where $\lambda = \exp\left[\theta_0 + \theta_1\left(1/(Z+273.16) - 1/(25+273.16)\right)\right]$, the reliability function is simply given by: $R(t;\underline{\Theta},Z) = \exp\left(-\exp\left[\theta_0 + \theta_1\left(1/(Z+273.16) - 1/(25+273.16)\right)\right]t\right)$ where Z is a constant stress and model parameter vector $\underline{\Theta} = [\theta_0, \theta_1]$. The baseline ML estimate of $\underline{\Theta}$ from a pilot experiment is: $\hat{\underline{\Theta}} = [\hat{\theta}_0, \hat{\theta}_1] = [-7.36, -339.80]$. The goal is to plan an ALT experiment that consumes the least amount of energy while providing the most precise estimate of reliability $\hat{R}(t_0; \hat{\underline{\Theta}}, Z_0)$ at the mission time $t_0 = 1000$ hours under the normal operating condition $Z_0 = 25°C$. The resources and constraints are as follows: (1) $n_{total} = 100$ test units are available; (2) the amplitude of the stress loading $75°C \le Z_i \le 150°C$, meanwhile the ramp rate $|dZ_i / dt| \le 2°C / \min$; (3) the test must be completed in 200 hours. In particular, we consider a scenario where sequential optimization has the flexibility of choosing controllers by taking advantage of equivalent ALT plans (Liao and Elayed, 2010). Suppose the test equipment is a first-order system with transfer function: $H(s) = 38.8/(2475s + 1)$ (Yang et al., 2006), a PID controller is used to control the equipment, and $\mathbb{P}(u(t; Z_i))$ is a quadratic function of $u(t; Z_i)$:

$\mathbb{P}(u(t; Z_i)) = 1 + u^2(t, Z_i)$. Depending on whether or not the PID controller is adjustable, different cases are studied next.

5.1.1 Test Equipment with an Adjustable PID Controller

When the three control parameters K_p, K_i and K_d are adjustable, these parameters are adjusted to make $\tilde{Z}_i$ follow Z_i, although not always feasible considering all the constraints. Assuming the capability of the system is limited, two levels of *ramp-constant-stress* loading (see Table 5-1) are utilized in sequential optimization process for planning the statistically efficient ALT experiment given in Problem *P1* with the decision variables: $\Omega^{(0)} = [Z_1^{(0)}, Z_2^{(0)}, n_1^{(0)}, n_2^{(0)}]$. From the resulting ALT plan $[\Omega^{(0)*}, \tau^{(0)}]$, we find an optimal equivalent ALT plan using a *modified-step-stress* loading Z_1^* (see Table 5-1), which minimizes the total energy consumption for the entire test time τ^*. This type of stress loading is used intentionally to save the test time (Park and Yum, 1998). Since all $n_{total} = 100$ units are tested at the same stress loading in this case, the decision variables of the desired statistically and energy efficient ALT plan are: $\Omega^{(0)} = Z_1$, τ, and $[K_p, K_i, K_d]$ of the PID controller.

The mathematical formulation for this sequential ALT planning process consisting of two nonlinear programs is given by:

$P1$: **Statistically efficient ALT plan**

$$\Omega^{(0)*} = \underset{\Omega^{(0)}}{\arg\min} \, Asvar\left(\hat{R}(t_0 = 1000\text{hrs}; \hat{\underline{\Theta}}, 25^o C \,|\, [\Omega^{(0)}, \tau^{(0)} = 200\text{hrs}])\right) \qquad (5.1)$$

Subject to
$$n_1 + n_2 = n_{total} = 100, \quad n_1, n_2 \in \{1, 2, \ldots\},$$
$$75^o C \le Z_i \le 150^o C, \qquad i = 1, 2,$$
$$|dZ_i(t)/dt| \le 2^o C / \min, \quad i = 1, 2.$$

Take the solution $[\Omega^{(0)*}, \tau^{(0)}]$ as the input to $\rightarrow$ continued

$\rightarrow P2$: **Desired statistically and energy efficient ALT plan**

$$[Z_1^*, \tau^*, K_p^*, K_i^*, K_d^*] = \underset{Z_1, \tau, K_p, K_i, K_d}{\arg\min} \; \mathbb{EN}(\tau, \{u(t; Z_1), 0 \le t \le \tau\}) = \int_0^\tau 1 + u^2(t; Z_1) \, dt \qquad (5.2)$$

Subject to
$$Asvar\left(\hat{R}(t_0 = 1000\text{hrs}; \hat{\underline{\Theta}}, 25^o C, n_{total} = 100 \,|\, [Z_1, \tau])\right)$$
$$= Asvar\left(\hat{R}(t_0 = 1000\text{hrs}; \hat{\underline{\Theta}}, 25^o C \,|\, [\Omega^{(0)*}, \tau^{(0)}])\right),$$

$$\left.\begin{array}{l} 75^o C \le Z_1 \le 150^o C, \\[4pt] |dZ_1(t)/dt| \le 2^o C / \min, \\[4pt] 0 < \tau \le 200 \text{ hours}, \end{array}\right\} \text{same as those in } P1$$

$$\mathbb{L}_s\left(\tilde{Z}_1(t)\right) = \mathbb{L}_s\left(u(t; Z_1)\right) H(s), \; \text{(equipment's response to input)}$$

$$RMSE = \sqrt{\frac{\sum_{j=1}^{N}(Z_1(t_j) - \tilde{Z}_1(t_j))^2}{N}} \le 0.5^o C, \quad 0 \le t_j \le \tau,$$

$$u(t; Z_1) = K_p e(t) + K_i \int_0^t e(v)\, dv + K_d [de(t)/dt],$$

$$0.1 < K_p \leq 5,$$

$$0 < K_i \leq 1.5,$$

$$0 < K_d \leq 100,$$

where $\mathbb{L}_s(\cdot)$ denotes Laplace transform given by $\mathbb{L}_s(g(t)) = \int_0^\infty e^{-st} g(t)dt$ for a function

$g(t)$ with $t \geq 0$, and N is the number of sample points used for evaluating the tracking

error between the desired stress loading and the actual output of the equipment.

When the *ramp-constant-stress* loadings or the *modified-step-stress* loading are used

in ALT experiments, the corresponding probability density function $f(t; Z_i)$ of failure

times and reliability function $R(t; Z_i)$ under Z_i can be expressed as (Liao and Elayed,

2010; Hong *et al.* 2010):

$$f(t; Z_i) = \exp\left[\theta_0 + \theta_1\left(1/(Z_i(t) + 273.16) - 1/(25 + 273.16)\right)\right]$$
$$\times \exp\left[-\exp(\theta_0)\int_0^t \exp\left[\theta_1\left(1/(Z_i(u) + 273.16) - 1/(25 + 273.16)\right)\right]du\right], \tag{5.3}$$

$$R(t; Z_i) = \exp\left[-\exp(\theta_0)\int_0^t \exp\left[\theta_1\left(1/(Z_i(u) + 273.16) - 1/(25 + 273.16)\right)\right]du\right], \tag{5.4}$$

respectively. The associated log-likelihood function is given by:

$$L(\underline{\Theta}\,|\,\text{data}) = \sum_{i=1}^{m}\sum_{j=1}^{n_i}\left[\delta_{ij}\ln\left(f(t_{ij}; Z_i)\right) + (1 - \delta_{ij})\ln\left(R(\tau; Z_i)\right)\right], \tag{5.5}$$

where $m = 2$ for the *ramp-constant-stress* loadings, $m = 1$ for the *modified-step-stress*

loading, and $\delta_{ij} = \{1, \text{ if } t_{ij} < \tau; 0, \text{ otherwise}\}$. Then, the two quantities measuring the

estimation precision of the two ALT experiments with different stress loadings can be

calculated by:

$$Asvar\left(\hat{R}(1000\text{hrs};\hat{\underline{\Theta}},25^{o}C\,|\,[\Omega^{(0)},\tau^{(0)}])\right)=\left[\frac{\partial\hat{R}(1000\text{hrs};\hat{\underline{\Theta}},25^{o}C)}{\partial\hat{\underline{\Theta}}}\right]^{T}\hat{\Sigma}_{\hat{\underline{\Theta}}}^{[\Omega^{(0)},\tau^{(0)}]}\left[\frac{\partial\hat{R}(1000\text{hrs};\hat{\underline{\Theta}},25^{o}C)}{\partial\hat{\underline{\Theta}}}\right] \quad (5.6)$$

and

$$Asvar\left(\hat{R}(t_{0}=1000\text{ hrs};\hat{\underline{\Theta}},25^{o}C\,|\,[Z_{1},\tau,n_{total}])\right)$$
$$=\left[\frac{\partial\hat{R}(1000\text{ hrs};\hat{\underline{\Theta}},25^{o}C)}{\partial\hat{\underline{\Theta}}}\right]^{T}\hat{\Sigma}_{\hat{\underline{\Theta}}}^{[Z_{1},\tau,n_{total}]}\left[\frac{\partial\hat{R}(1000\text{ hrs};\hat{\underline{\Theta}},25^{o}C)}{\partial\hat{\underline{\Theta}}}\right] \quad (5.7)$$

according to Eq. (3.2) and using

$$\left[\frac{\partial\hat{R}(1000\text{ hrs};\hat{\underline{\Theta}},25^{o}C)}{\partial\hat{\underline{\Theta}}}\right]^{T}=\left[-1000\exp(\theta_{0})\exp\left(-1000\exp(\theta_{0})\right),0\right] \quad (5.8)$$

and the corresponding variance-covariance matrices $\hat{\Sigma}_{\hat{\underline{\Theta}}}^{[\cdot]}=\left[E\left[-\frac{\partial^{2}L(\Theta|\text{data})}{\partial\underline{\Theta}\partial\underline{\Theta}^{T}}\right]\right]^{-1}$ evaluated at

$\hat{\underline{\Theta}}$ (see Appendix A for the general expressions of the terms in $E\left[-\frac{\partial^{2}L(\Theta|\text{data})}{\partial\underline{\Theta}\partial\underline{\Theta}^{T}}\right]$ for the

ALT experiments using either the *ramp-constant-stress* loadings or the *modified-step-stress* loading).

Solving Problem *P1* yields: $Z_{1}^{(0)*}$, $Z_{2}^{(0)*}$ (see Table 5-1), $n_{1}^{(0)*}=84$, and $n_{2}^{(0)*}=16$. This is the traditional optimum ALT plan focusing only on statistical efficiency. When solving Problem *P2*, a stress loading Z_{1} and test time τ are obtained by solving the first equality based on the work of Liao and Elsayed (2010) considering the three inequality constraints same as those in Problem *P1*; then a PID controller is fine-tuned after using the Ziegler–Nichols rule in Matlab Simulink to make $\tilde{Z}_{1}$ follow Z_{1} within a specified limit for tracking error (RMSE $\leq 0.5^{o}C$ in this example); if feasible, the energy consumption associated with Z_{1} is calculated up to time τ ; otherwise, this ALT plan is eliminated and this process is repeated to generate another pair of $[Z_{1},\tau]$; finally, the desired ALT plan $[Z_1^*, \tau^*]$ along with the controller parameters $[K_p^*, K_i^*, K_d^*]$ consuming the least amount of energy is selected from all the feasible equivalent ALT plans.

Table 5-1 presents the resulting ALT plans, the associated controllers, and the corresponding values of energy consumption. Compared to the benchmark ALT plan targeting only on the estimation precision, the desired ALT plan reduces the energy consumption by 61.7% without sacrificing the estimation precision. This is a big percentage energy savings in ALT.

Table 5-1: Results for the statistically efficient ALT plan and the desired ALT plan (C1)

ALT plans	Stress loadings and other decision variables	Parameters of PID controller, control signal $u(t; Z_i)$, and total energy use
Benchmark statistically efficient ALT plan $Asvar\left(\hat{R}(1000 \text{ hrs};25^{\circ}C)\right)$ $= 0.0647$	$n_1^{(0)*} = 84, n_2^{(0)*} = 16,$ $\tau^{(0)} = 200$ hours. **Two levels of ramp-constant-stress loading**	$K_p^* = 4.9904,$ $K_i^* = 0.75,$ $K_d^* = 0.01.$ Total energy use $=15,572,957$ units of energy
Desired statistically & energy efficient ALT plan $Asvar\left(\hat{R}(1000 \text{ hrs};25^{\circ}C)\right)$ $= 0.0647$	$n_{total} = 100,$ $\tau^* = 199.4$ hours. **One modified-step-stress loading**	$K_p^* = 4.9904,$ $K_i^* = 0.75,$ $K_d^* = 0.01.$ Total energy use $=5,968,637$ units of energy

5.1.2 Test Equipment with an Unadjustable PID Controller

When parameters K_p, K_i and K_d of the given PID controller are fixed, an input-weighting PID controller (see Figure 3-7) can be used to improve the performance of the test equipment. As a result, the control signal to the equipment can be expressed as:

$$
\begin{aligned}
u(t;Z_i) = K_p\left[F_p Z_i(t) - \tilde{Z}_i(t)\right] + K_i\int_0^t \left[F_i Z_i(t) - \tilde{Z}_i(t)\right] dv \\
+ K_d\left[d\left(F_d Z_i(t) - \tilde{Z}_i(t)\right)/dt\right],
\end{aligned}
\tag{5.9}
$$

where $F_i = 1$. Instead of adjusting the values of these three PID control parameters, the values of F_p and F_d need to be tuned in such a way that the energy consumption and the tracking error of the test equipment can be minimized while satisfying other constraints. Due to the limitation of the test equipment and the features of the adopted stress-loading profile, the desired input-weighting PID controller is not always feasible. Similar to the previous case, two levels of *ramp-constant stress* loading are first utilized in planning the statistically efficient ALT experiment given in Problem *P1* with the decision variables: $\Omega^{(0)} = [Z_1^{(0)}, Z_2^{(0)}, n_1^{(0)}, n_2^{(0)}]$. Based on the resulting statistically efficient ALT plan $[\Omega^{(0)*}, \tau^{(0)}]$, an equivalent ALT plan using a *modified-step-stress* Z_1^* can be found, which consumes the least amount of energy in the entire test time τ^* . The decision variables of the desired statistically and energy efficient ALT plan are: $[Z_1, \tau]$ and $[F_p, F_d]$.

The mathematical formulation for this sequential ALT planning process is given by:

P1:   **Statistically efficient ALT plan**

$$
[\Omega^{(0)*}, \tau^{(0)}] = \underset{\Omega^{(0)}}{\arg\min} Asvar\left(\hat{R}(t_0 = 1000\text{hrs}; \hat{\Theta}, 25^\circ C \mid [\Omega^{(0)}, \tau^{(0)} = 200\text{hrs}])\right)
\tag{5.10}
$$

Subject to $\quad n_1 + n_2 = n_{total} = 100, \ n_1, n_2 \in \{1, 2, ...\},$

$\qquad\qquad 75^{\circ}C \le Z_i \le 150^{\circ}C, \qquad i = 1, 2,$

$\qquad\qquad |dZ_i(t)/dt| \le 2^{\circ}C/\min, \quad i = 1, 2.$

Take the solution $[\Omega^{(0)*}, \tau^{(0)}]$ as the input to $\rightarrow$ continued

$\rightarrow P2$: Desired statistically and energy efficient ALT plan

$$[Z_1^*, \tau^*, F_p^*, F_d^*] = \underset{Z_1, \tau, F_p, F_d}{\arg\min} \ \mathbb{EN}(\tau, \{u(t; Z_1), 0 \le t \le \tau\}) = \int_0^{\tau} 1 + u^2(t; Z_1)\, dt \qquad (5.11)$$

Subject to $\quad Asvar\left(\hat{R}(t_0 = 1000\mathrm{hrs}; \hat{\Theta}, 25^{\circ}C, n_{total} = 100 \,|\, [Z_1, \tau])\right)$

$$= Asvar\left(\hat{R}(t_0 = 1000\mathrm{hrs}; \hat{\Theta}, 25^{\circ}C \,|\, [\Omega^{(0)*}, \tau^{(0)}])\right),$$

$$\left.\begin{array}{l} 75^{\circ}C \le Z_1 \le 150^{\circ}C, \\[4pt] |dZ_1(t)/dt| \le 2^{\circ}C/\min, \\[4pt] 0 < \tau \le 200 \text{ hours,} \end{array}\right\} \text{same as those in } P1$$

$$\mathbb{L}_s\left(\tilde{Z}_1(t)\right) = \mathbb{L}_s\left(u(t; Z_1)\right) H(s), \ \text{(equipment's response to input)}$$

$$RMSE = \sqrt{\frac{\sum_{j=1}^{N}(Z_1(t_j) - \tilde{Z}_1(t_j))^2}{N}} \le 0.5^{\circ}C, \quad 0 \le t_j \le \tau,$$

$$u(t; Z_1) = K_p\left(F_p Z_1(t) - \tilde{Z}_1(t)\right) + K_i \int_0^t \left(Z_1(v) - \tilde{Z}_1(v)\right) dv$$
$$+ K_d\left[d\left(F_d Z_1(t) - \tilde{Z}_1(t)\right)/dt\right],$$

$$0.05 < F_p \le 1.5,$$
$$1 < F_d \le 7000.$$

Again, by solving the first equality in Eq. (5.11) while satisfying the three inequality constraints included in Problem $P1$, a stress loading Z_1 and test time τ are obtained. Since the PID controller of the system is fixed, for the resulting stress loading Z_1 and test time τ, the original closed-loop system could be either under-damped or over-damped (see Figure 5-1 and Figure 5-2). Next, we study these two different subcases.

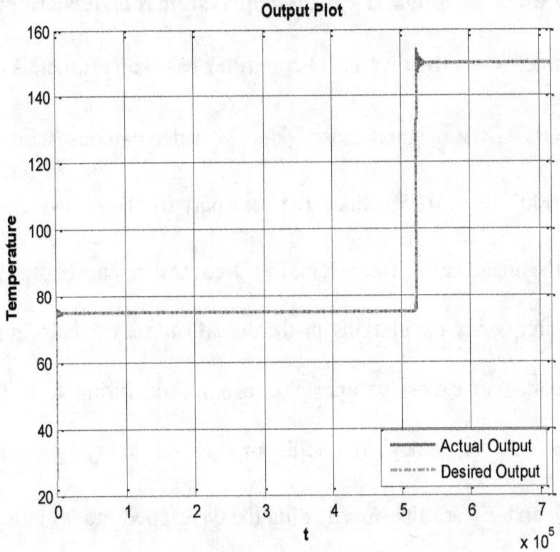

Figure 5-1: Original closed-loop system response: under-damped

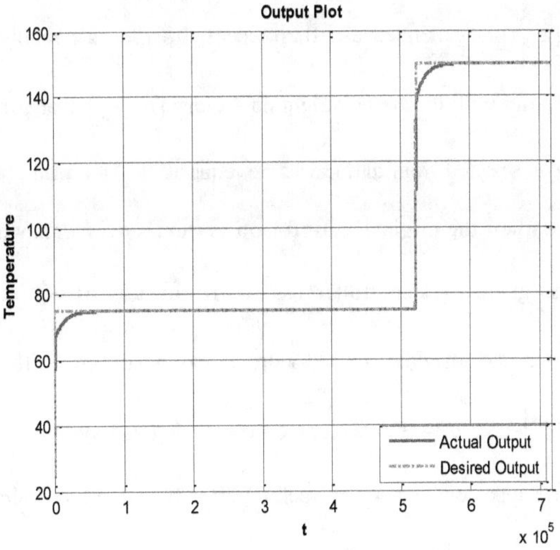

Figure 5-2: Original closed-loop system response: over-damped

(1) Case when the original closed-loop system is under-damped:

For example, when the given PID controller has the parameters of $K_p = 0.01$, $K_i = 0.01$ and $K_d = 0.01$, the original closed-loop is under-damped with a pair of complex poles $p_{1,2} = -0.0003 \pm 0.0125i$. Since the real part of these two closed-loop poles are very close to the image axis, the original PID controller can cause large amplitude overshot and high frequency oscillations in the actual output of the original system, which often closely related to excessive energy consumption during test. To reduce the overshoot amplitude and frequency of oscillations exists in system response, input-weighting factors F_p and F_d are fine-tuned using the developed model to make $\tilde{Z}_1$ follow Z_1 within a specified tracking error range (RMSE $\leq 0.5^\circ C$); if feasible, the energy consumption associated with Z_1 is calculated up to time τ ; otherwise, this ALT plan is eliminated and another $[Z_1, \tau]$ is generated and then repeat this process; finally, the desired ALT plan $[Z_1^*, \tau^*]$ along with the input-weighting factors $[F_p^*, F_d^*]$ that consumes the least amount of energy is selected from all the feasible equivalent ALT plans.

(2) Case when the original closed-loop system is over-damped:

When the given PID controller has the parameters of $K_p = 0.11$, $K_i = 0.00001$ and $K_d = 0.01$, the original closed-loop is over-damped with a pair of real poles $p_1 = -0.0001$ and $p_2 = -0.0021$, which cause sluggish system response. Moreover, these values may also result in large tracking error between the desired and the actual stress conditions. By manipulating the values of F_p and F_d, the input-weighting PID controller can accelerate system response speed and therefore reduce the tracking error to an acceptable range (RMSE $\leq 0.5^{\circ}C$). If the desired input-weighting PID controller is feasible for the current plan $[Z_1, \tau]$, the energy consumption associated with Z_1 is calculated until time τ; otherwise, this ALT plan is excluded and another $[Z_1, \tau]$ will be generated to repeat the previous process. In the end, by comparing the energy consumption of all candidate plans, the desired ALT plan $[Z_1^*, \tau^*]$ along with the input-weighting factors $[F_p^*, F_d^*]$ consuming the least amount of energy is selected to be the statistically and energy efficient ALT plan.

Table 5-2 summarizes the statistically efficient ALT plan and the desired ALT plan obtained in scenario 2 under two different situations (over-damped and under-damped). The associated controllers and the corresponding values of energy consumption are also listed in this table.

Table 5-2: Results for the statistically efficient ALT plan and the desired ALT plan (C2)

ALT plans	Stress loadings and other decision variables	Parameters of PID controller, control signal $u(t; Z_i)$, and total energy use
Benchmark statistically efficient ALT plan $Asvar\left(\hat{R}(1000 \text{ hrs}; 25^{\circ}C)\right)$ = 0.0647	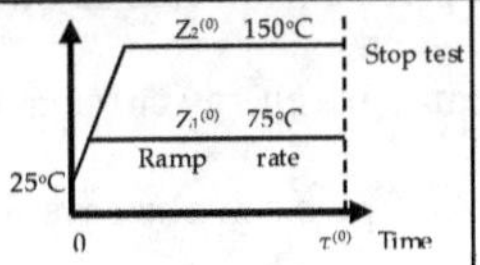 $n_1^{(0)*} = 84, n_2^{(0)*} = 16,$ $\tau^{(0)} = 200$ hours.	**Under-damped** K_p=0.01, K_i=0.01, K_d=0.01 Total energy use = 15,368,730 units of energy 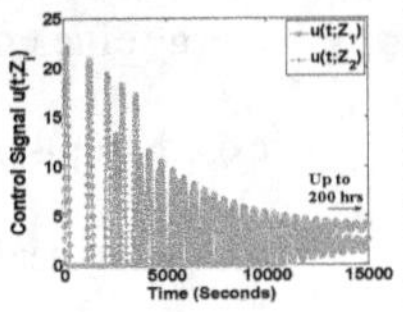

	Two levels of ramp-constant-stress loading	Over-damped
		K_p=0.11, K_i=0.00001, K_d=0.01. Total energy use =14,842,210 units of energy
Desired statistically & energy efficient ALT plan $Asvar\left(\hat{R}(1000\ \text{hrs};25^\circ C)\right) = 0.0647$	$n_{total} = 100,$ $\tau^* = 199.4$ hours. **One modified-step-stress loading**	**Under-damped** $F_p^* = 0.10,$ $F_d^* = 6387.25$ Total energy use =5,643,278 units of energy **Over-damped** $F_p^* = 1.17,$ $F_d^* = 5850.03$ Total energy use =5,627,437 units of energy

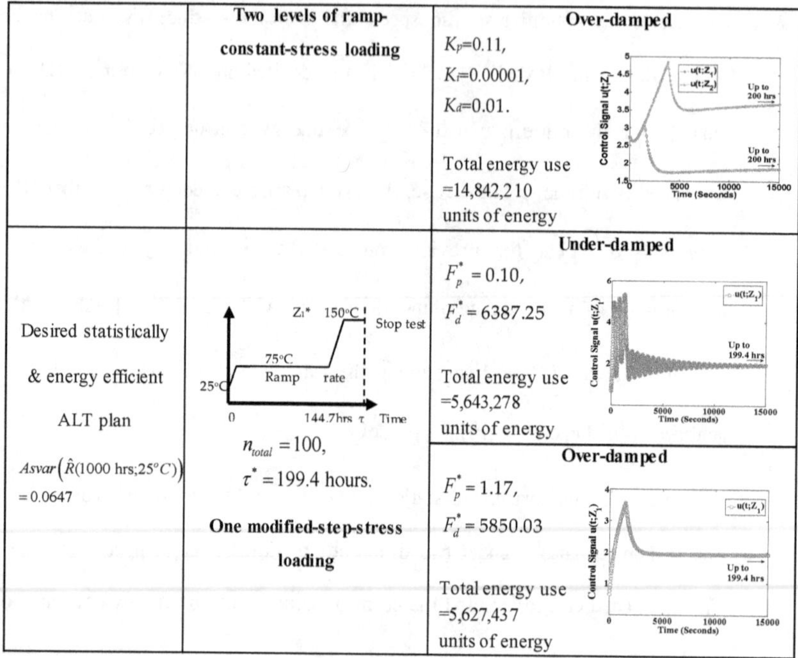

As seen on Table 5-2, when PID controller is unadjustable, the proposed method can efficiently reduce the energy consumption for both under-damped and over-damped system without sacrificing the precision of estimation. The desired statistically and energy efficient ALT plan can reduce the energy consumption by 62.1% for over-damped closed-loop system; the desired ALT plan can also save 63.3% in energy uses for under-damped system.

5.2 Application of Simultaneous Optimization ALT Design Method

In this case study, we assume the product's failure time under a constant stress follows a

Weibull distribution with cumulative distribution function (cdf) and probability density function (pdf):

$$F(t;\eta,\beta) = 1 - e^{-(t/\eta)^{\beta}} , \qquad (5.12)$$

$$f(t;\eta,\beta) = \frac{\beta}{\eta}\left(\frac{t}{\eta}\right)^{\beta-1} e^{-(t/\eta)^{\beta}} , \qquad (5.13)$$

where η is the scale parameter and β is the shape parameter. Weibull distribution belongs to the log-location-scale distribution family and is a widely used in reliability engineering. Let $\mu = \ln(\eta)$ and $\sigma = 1/\beta$. We can express the cdf and pdf of the product's failure time as:

$$F(t;\mu,\sigma) = \Phi_{sev}\left(\frac{\ln(t)-\mu}{\sigma}\right), \qquad (5.14)$$

$$f(t;\mu,\sigma) = \frac{1}{\sigma t}\phi_{sev}\left(\frac{\ln(t)-\mu}{\sigma}\right) \qquad (5.15)$$

where $\Phi_{sev}(x) = 1 - \exp(-\exp(x))$ and $\phi_{sev}(x) = \exp(x - \exp(x))$ are the cdf and pdf of the standard smallest extreme value distribution, respectively. The ALT model is assumed to have a log-linear life-stress relationship where

$$\mu = g(Z;\underline{\theta}) = \theta_0 + \theta_1\left[1/(Z_0 + 273.16) - 1/(Z + 273.16)\right]. \qquad (5.16)$$

In Eq. (5.16), Z_0 represents normal operation condition of the product and Z refers to the stress level of interest. The scale parameter $\sigma = 1/\beta$ is assumed to remain constant across the range of stresses considered. The baseline ML estimate of $\underline{\Theta}$ is: $\hat{\underline{\Theta}} = [\hat{\theta}_0, \hat{\theta}_1] = [8.7519, -3906.86]$ and $\hat{\sigma} = 0.5$. The goal is to find a simple SSALT plan that consumes the least amount of energy while providing the most precise estimate of natural logarithm of the 0.1 quantile under the use stress level $Z_0 = 25°C$, where

$$\hat{y}_{0.1}(\hat{\Theta}, Z_0) = \log(t_{0.1}; \hat{\Theta}, Z_0) = \hat{\theta}_0 + \hat{\theta}_1 \left[1/(Z_0 + 273.16) - 1/(Z + 273.16) \right] + \Phi_{sev}^{-1}(0.1)\hat{\sigma} \,. (5.17)$$

In Eq. (5.17), $\Phi_{sev}^{-1}(0.1) = \ln\left[-\ln(1 - 0.1) \right]$ is the 0.1 quantile of standard smallest extreme value distribution.

The resources and constraints in this case study are listed as follows:

(1) $n_{total} = 50$ test units are available;

(2) The amplitude of the stress loading $25°C \leq Z_i \leq 150°C$;

(3) The test must be completed in 380 hours.

Similar to the case study provided in Section 5.1, in the simultaneous optimization case we also assume the test equipment is a first-order system with transfer function: $H(s) = 38.8/(2475s + 1)$, which is controlled by a PID controller and the instantaneous power consumption $\mathbb{P}(u(t; Z_i))$ can be estimated using a quadratic function of $u(t; Z_i)$: $\mathbb{P}(u(t; Z_i)) = 1 + u^2(t, Z_i)$. Again, depending on whether or not the PID controller is adjustable, the simultaneous optimization algorithms developed in Sections 4.2.2 and 4.2.3 are implemented to find statistically and energy efficient ALT plans, respectively.

5.2.1 Test Equipment with an Adjustable PID Controller

For test equipment with an adjustable PID controller, the three control parameters K_p, K_i and K_d can be adjusted to generate control signals that make the actual response of the test equipment $\tilde{Z}_i$ follow a given stress loading profile Z_i. As one of the most widely used stress loadings in reliability testing, *simple step-stress* loading are utilized in planning the statistically and energy efficient ALT experiment using the simultaneous optimization ALT design procedure proposed in Section 4.2.1. For *simple step-stress* all $n_{total} = 50$ units are tested at the same stress loading simultaneously, so the vector of decision variables of the desired ALT plan is: $\underline{X}=[\underline{P},\tau,K_p,K_i,K_d]$, where $\underline{P}$ represents the *simple step-stress* profile adopted in the ALT plan which consists of first step level Z_1, second step level Z_2, and step change time τ_1. τ is the censoring time of designed ALT plan.

Given the resources and constraints stated previously, the mathematical formulation of this fully integrated ALT planning process can be represented as:

$$\underline{X}^* = \operatorname*{arg\,min}_{\underline{X}=[\underline{P},\tau,K_p,K_i,K_d]} \{f_1(\underline{X}), f_2(\underline{X}), f_3(\underline{X})\} \tag{5.18}$$

$$\text{Subject to} \quad 25^{\circ}C \le Z_i \le 150^{\circ}C, i = \{1,2\},$$

$$0 < \tau_1 \le 380 \text{ hours},$$

$$\tau_1 < \tau \le 380 \text{ hours},$$

$$\mathbb{L}_s\left(\underline{\tilde{P}}(t)\right) = \mathbb{L}_s\left(u(t;\underline{P})\right)H(s), \quad \text{(equipment's response to input)}$$

$$u(t;\underline{P}) = K_p e(t) + K_i \int_0^t e(v)\,dv + K_d[de(t)/dt],$$

$$0.1 < K_p \le 5,$$

$$0 < K_i \le 1.5,$$

$$0 < K_d \le 100,$$

where

$$f_1(\underline{X}) = \mathbb{EN}(\tau, \{u(t;\underline{P}), 0 \le t \le \tau\}) = \int_0^\tau 1 + u^2(t;\underline{P})\,dt \;, \tag{5.19}$$

$$f_2(\underline{X}) = Asvar\left(\hat{y}_{0.1}(\hat{\underline{\Theta}}, Z_0, n_{total} = 50 \mid [\underline{P}, \tau, K_p, K_i, K_d])\right), \qquad (5.20)$$

$$f_3(\underline{X}) = RMSE = \sqrt{\frac{\sum_{j=1}^{N}\left[\underline{P}(t_j) - \hat{\underline{P}}(t_j)\right]^2}{N}}, \quad 0 \le t_j \le \tau. \qquad (5.21)$$

When the *step-stress* loading is used in ALT experiment, based on Khamis-Higgins model (Khamis and Higgins, 1998) the corresponding cdf $F(t;\underline{P})$ and pdf $f(t;\underline{P})$ of failure times under $\underline{P}$ can be expressed as:

$$F(t;\underline{P}) = \begin{cases} 1 - \exp\left[-\exp(\dfrac{\ln(t) - \mu_1}{\sigma})\right] & 0 \le t < \tau_1 \\[3mm] 1 - \exp\left[-\exp(\dfrac{\ln(t) - \mu_2}{\sigma}) + \sum_{i=1}^{2}(-1)^i \exp(\dfrac{\ln(\tau_1) - \mu_i}{\sigma})\right] & \tau_1 \le t < \infty \end{cases}, \qquad (5.22)$$

$$f(t;\underline{P}) = \begin{cases} \dfrac{1}{\sigma t}\exp\left[\dfrac{\ln(t) - \mu_1}{\sigma} - \exp(\dfrac{\ln(t) - \mu_1}{\sigma})\right] & 0 \le t < \tau_1 \\[3mm] \dfrac{1}{\sigma t}\exp\left[\dfrac{\ln(t) - \mu_2}{\sigma} - \exp(\dfrac{\ln(t) - \mu_2}{\sigma}) + \sum_{i=1}^{2}(-1)^i \exp(\dfrac{\ln(\tau_1) - \mu_i}{\sigma})\right] & \tau_1 \le t < \infty \end{cases}, \qquad (5.23)$$

where $\mu_i = \theta_0 + \theta_1\left[1/(Z_0 + 273.16) - 1/(Z_i + 273.16)\right], i = \{1, 2\}$.

Correspondingly, the reliability function under $\underline{P}$ can be expressed as:

$$R(t;\underline{P}) = \begin{cases} \exp\left[-\exp(\dfrac{\ln(t) - \mu_1}{\sigma})\right] & 0 \le t < \tau_1 \\[3mm] \exp\left[-\exp(\dfrac{\ln(t) - \mu_2}{\sigma}) + \sum_{i=1}^{2}(-1)^i \exp(\dfrac{\ln(\tau_1) - \mu_i}{\sigma})\right] & \tau_1 \le t < \infty \end{cases}. \qquad (5.24)$$

The associated log-likelihood function is given by:

$$L(\underline{\Theta} \mid \text{data}) = \sum_{j=1}^{n}\left[\delta_j \ln\left(f(t_j;\underline{P})\right) + (1 - \delta_j)\ln\left(R(\tau;\underline{P})\right)\right], \qquad (5.25)$$

where $n = 50$ in this example and $\delta_j = \{1, \text{ if } t_j < \tau; 0, \text{ otherwise}\}$. The estimation precision of the designed ALT experiments can be quantified by:

$$Asvar\left(\hat{y}_{0.01}(\hat{\underline{\Theta}}, 25^{o}C, n_{total} = 50\,|\,[\underline{P}, \tau, K_{p}, K_{i}, K_{d}])\right) = \left[\frac{\partial \hat{y}_{0.01}(\hat{\underline{\Theta}}, 25^{o}C)}{\partial \hat{\underline{\Theta}}}\right]^{T} \hat{\Sigma}_{\hat{\underline{\Theta}}}^{[\underline{X}]} \left[\frac{\partial \hat{y}_{0.01}(\hat{\underline{\Theta}}, 25^{o}C)}{\partial \hat{\underline{\Theta}}}\right].\,(5.26)$$

Based on Eq. (5.17), we have $\left[\frac{\partial \hat{y}_{0.01}(\hat{\underline{\Theta}}, 25^{o}C)}{\partial \hat{\underline{\Theta}}}\right]^{T} = \left[1, 0, \Phi_{sev}^{-1}(0.1)\right]$. The corresponding

variance-covariance matrix $\hat{\Sigma}_{\hat{\underline{\Theta}}}^{[\underline{X}]} = \left[E\left[-\frac{\partial^{2}L(\underline{\Theta}|\text{data})}{\partial \underline{\Theta} \partial \underline{\Theta}^{T}}\right]\right]^{-1}$ evaluated at $\hat{\underline{\Theta}}$ can be estimated

using the terms in $E\left[-\frac{\partial^{2}L(\underline{\Theta}|\text{data})}{\partial \underline{\Theta} \partial \underline{\Theta}^{T}}\right]$ provided in Appendix B. Appendix B also provides the

first and second partial derivatives of Eq. (5.25) with respect to the model parameters for

the ALT experiments using the *simple step-stress* loading.

In order to obtain the statistically and energy efficient ALT plan, the developed

simultaneous optimization algorithms introduced in Section 4.2.1 is utilized to solve

Problem (5.18). At the end of this optimization process, 37 Pareto optimal solutions are

obtained and their corresponding objective values are shown in Table 5-3. Figure 5-3

displays these solutions in a 3-D space.

Table 5-3: Objective values of resulting Pareto optimal solutions (Adjustable PID)

Solution #	$f_1(\underline{X})$ $(\times 10^7)$	$f_2(\underline{X})$	$f_3(\underline{X})$	Solution #	$f_1(\underline{X})$ $(\times 10^7)$	$f_2(\underline{X})$	$f_3(\underline{X})$
1	1.5416	0.5323	0.6578	20	1.3973	0.6171	0.8573
2	0.2653	27.6019	0.1584	21	0.4545	11.0164	0.1580
3	0.2546	33.4610	0.1537	22	0.7746	2.4980	0.2715
4	0.4197	21.6984	0.1117	23	0.3899	20.2237	0.1169
5	0.2442	39.0834	0.1551	24	0.4001	24.1306	0.1144
6	0.2405	36.3395	0.2418	25	2.8077	0.5679	0.3555
7	1.5086	0.5601	0.8412	26	0.6618	4.3347	0.1914
8	0.2544	35.9793	0.1741	27	1.8704	0.5335	0.5650
9	1.0582	1.2478	0.7834	28	0.3288	13.3735	0.1513
10	1.7437	0.5714	0.5056	29	0.4364	8.1398	0.1812
11	2.4651	0.5330	0.4036	30	1.3987	0.6310	0.6864

12	0.9692	1.4014	0.4643	31	2.583	0.7315	0.2784
13	1.1176	0.8873	0.4847	32	0.4925	7.0093	0.1799
14	1.2882	2.1555	0.3325	33	0.5535	6.2258	0.2524
15	0.445	11.7666	0.1561	34	0.4203	14.6429	0.1378
16	0.2831	25.6685	0.1369	35	2.8011	0.5693	0.2564
17	0.2448	39.0834	0.1534	36	0.2699	27.5999	0.1439
18	2.3436	0.5719	0.4371	37	3.0549	0.5349	0.3604
19	0.2955	16.8991	0.1429				

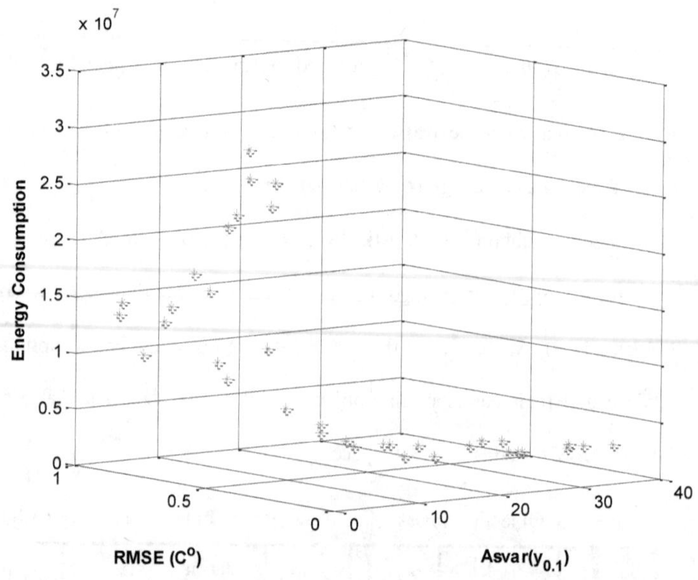

Figure 5-3: Objective values of Pareto optimal solutions in a 3-D space

For designing statistically and energy efficient ALT plans, our primary goal is to minimize the energy usage during test while providing accurate reliability estimation, so the first two objectives are of higher preference over the third objective. Considering capability limit of most test equipment, as long as the RMSE is below 1 ^{o}C, the tracking performance can be regarded as acceptable. Under this condition, it is reasonable to assume that the difference between designed and actual stress profile created by test equipment won't result in obvious changes in products failure process and in the subsequent reliability estimation. As shown in Table 5-3, the RMSEs of all Pareto optimal solutions are smaller than $1\,^{o}C$. Hence, in this example we can treat the third objective as a pseudo constraint and plot the values of $f_1(\underline{X})$ and $f_2(\underline{X})$ in a 2-D space (see Figure 5-4).

It can be clearly observed in Figure 5-4 that the proposed simultaneous optimization procedure can effectively search inside the design region for statistically and energy efficient ALT plans considering the trade-offs between energy consumption and reliability estimation precision. Based on the values of three objective functions displayed in Table 5-3 and the limitation of test equipment, we selected three the optimum ALT plans regarding individual optimization criteria and one overall optimum ALT plan considering the trade-offs among three objectives (see Table 5-4).

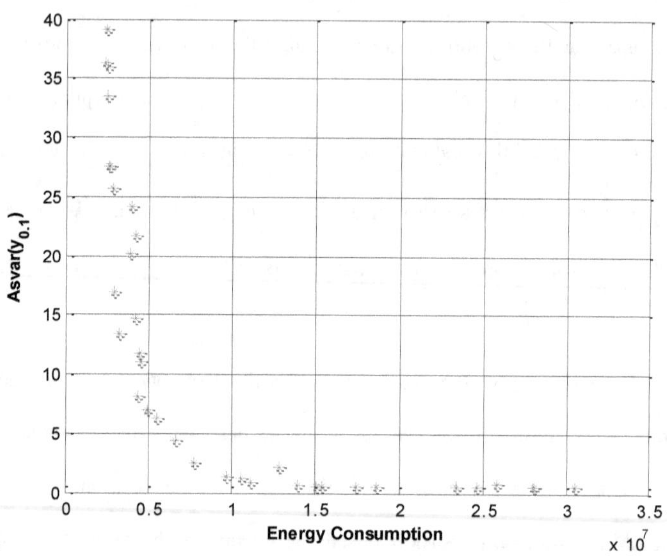

Figure 5-4: Pareto optimal solutions with the first two Objective values

Table 5-4: Selected optimum ALT plans (Adjustable PID)

Solution #	$\underline{X} = \left[Z_1, Z_2, \tau_1, \tau, K_p, K_I, K_d\right]$	$f_1(\underline{X})$	$f_2(\underline{X})$	$f_3(\underline{X})$	Note
6	[26, 76, 289.0, 304.2, 2.0672, 0.04, 13.0156]	2,404,918	36.3395	0.2418	Minimal energy consumption
1*	[97, 150, 361.2, 380.0, 4.9986, 0.62, 7.6678]	15,415,517	0.5323	0.6578	Optimal estimate accuracy
4	[32, 74, 290.9, 352.0, 4.9986, 0.03, 31.3101]	4,196,901	21.6984	0.1117	Minimal RMSE
1*	[97, 150, 361.2, 380.0, 4.9986, 0.62, 7.6678]	15,415,517	0.5323	0.6578	Overall optimum ALT

As seen in Table 5-4, Solution #6 minimizes the total energy consumption during test but significantly increase the uncertainty in reliability estimation. For Solution #4, optimum tracking performance of test equipment is achieved by sacrificing statistical efficiency of the corresponding ALT experiment. Given trade-offs among energy consumption, estimation precision and tracking performance, the overall optimum ALT plan designed using simultaneous optimization method is:

$$\underline{X}^* = \underline{X}^{(1)} = \left[Z_1^*, Z_2^*, \tau_1^*, \tau^*, K_p^*, K_i^*, K_d^* \right]$$
$$= [97, \ 150, \ 361.2, \ 380.0, \ 4.9986, \ 0.62, \ 7.6678] \tag{5.27}$$

For this plan, $Asvar\left(\hat{y}_{0.1}(\hat{\Theta}, Z_0, n_{total} = 50 \mid \underline{X}^*)\right) = 0.5323$ and RMSE is equal to 0.6578 which is smaller than $1\,^oC$. The estimated energy consumption of this ALT plan is 15,415,517 units of energy.

5.2.2 Test Equipment with an Unadjustable PID Controller

As discussed in section 4.4.2, when we are facing a piece of test equipment with an unadjustable PID controller, energy reduction through the adjustment of its PID control parameters is no longer a feasible solution. Hence, an input-weighting PID controller needs to be introduced to the original control system to improve the performance of the test equipment according to different stress profiles.

In this case study, we assume the given PID controller has the parameters of $K_p = 0.11$, $K_i = 0.001$ and $K_d = 0.01$ and our goal is to design statistically and energy efficient ALT experiments with *simple step-stress* loadings. Based on Eq. (4.7), we can formulate this simultaneous optimum ALT design problem as:

$$\underline{X}^* = \underset{X=[\tilde{P}, \tau, F_p, F_d]}{\arg \min} \ \{f_1(\underline{X}), f_2(\underline{X}), f_3(\underline{X})\} \tag{5.28}$$

Subject to $\quad 25^{\circ}C \leq Z_i \leq 150^{\circ}C, i=\{1,2\},$

$$10 < \tau_1 \leq 380 \text{ hours},$$

$$\tau_1 < \tau \leq 380 \text{ hours},$$

$$\mathbb{L}_s\left(\tilde{\underline{P}}(t)\right) = \mathbb{L}_s\left(u(t;\underline{P})\right)H(s), \text{ (equipment's response to input)}$$

$$u(t;Z_i) = K_p\left[F_pZ_i(t) - \tilde{Z}_i(t)\right] + K_i\int_0^t\left[F_iZ_i(t) - \tilde{Z}_i(t)\right]dv$$
$$+ K_d\left[d\left(F_dZ_i(t) - \tilde{Z}_i(t)\right)/dt\right],$$

$$0.05 < F_p \leq 1.5,$$

$$1 < F_d \leq 7000,$$

where

$$f_1(\underline{X}) = \mathbb{EN}(\tau, \{u(t;\underline{P}), 0 \leq t \leq \tau\}) = \int_0^\tau 1 + u^2(t;\underline{P})\, dt \ , \qquad (5.29)$$

$$f_2(\underline{X}) = Asvar\left(\hat{y}_{0.1}(\hat{\Theta}, Z_0, n_{total} = 50 | [\underline{P}, \tau, F_p, F_d])\right), \qquad (5.30)$$

$$f_3(\underline{X}) = RMSE = \sqrt{\frac{\sum_{j=1}^{N}\left[\underline{P}(t_j) - \tilde{\underline{P}}(t_j)\right]^2}{N}}, 0 \leq t_j \leq \tau . \qquad (5.31)$$

By solving this problem using the simultaneous optimization algorithm provided in Section 4.2.3, 32 Pareto optimal solutions are obtained and their corresponding objective values are shown in Table 5-5 and Figure 5-5.

Table 5-5: Objective values of resulting Pareto optimal solutions (Unadjustable PID)

Solution #	$f_1(\underline{X})$ $(\times 10^7)$	$f_2(\underline{X})$	$f_3(\underline{X})$	Solution #	$f_1(\underline{X})$ $(\times 10^7)$	$f_2(\underline{X})$	$f_3(\underline{X})$
1	0.1008	1419.3325	1.8085	17	0.2204	108.7045	1.2046
2	4.0252	0.6420	0.0169	18	2.8434	1.5892	0.2332
3	1.8225	0.5432	0.6275	19	0.7818	2.7516	0.8272
4	0.1164	855.8826	1.6475	20	2.6131	1.8786	0.2413
5	2.5991	1.4950	0.2505	21	0.1401	342.3838	1.9363
6	0.1149	746.0229	1.6330	22	0.1117	821.5858	1.8232
7	4.0255	0.6413	0.0436	23	4.0256	0.6417	0.0350

8	0.1045	997.1127	1.6822	24	0.1052	1050.1141	1.7474
9	0.3083	22.8651	1.0702	25	1.5195	3.4923	0.5165
10	0.1213	592.9853	2.0985	26	0.1045	1249.5283	1.8420
11	0.1588	203.7638	1.4252	27	1.1801	10.3572	0.5380
12	0.1812	174.1182	2.4442	28	0.1234	557.6557	1.9162
13	3.5351	1.5665	0.1050	29	3.2902	1.1498	0.1215
14	0.1806	172.7216	2.3129	30	0.1064	1108.2623	1.8250
15	1.8268	2.4436	0.3066	31	0.1907	155.8999	2.6702
16	0.0998	1378.9418	1.8741	32	0.0995	1388.1387	1.8760

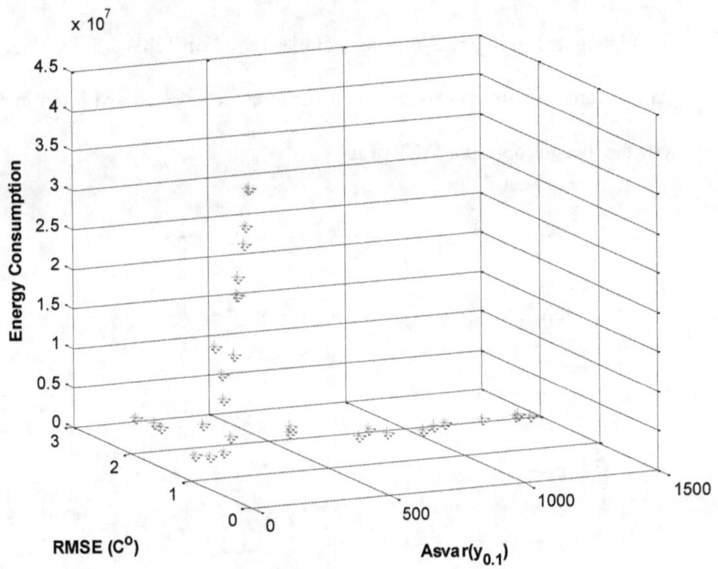

Figure 5-5: Objective values of Pareto optimal solutions in a 3-D space

As shown in Table 5-5 and Figure 5-5, the energy consumption value of the obtained Pareto optimum ALT plans varies from 995,431 to 40,255,801 units of energy; the asymptotic variance of natural logarithm of the 0.1 quantile under the use stress level varies from 0.5432 to 1419.3325; the RMSE between the desired and actual test condition ranges from 0.0169 to 2.6702 ^{o}C. Among all alternative optimal solutions, 19 solutions have RMSE values greater than 1 ^{o}C. After removing these 19 solutions, the values of energy consumption and $As\,\mathrm{var}(\hat{y}_{0.1})$ of the rest of feasible Pareto optimal solutions are plotted in Figure 5-6. As shown in Figure 5-6, the energy consumption of ALT experiment reduces as the estimation uncertainty increases. An overall optimal ALT plan with the desired energy consumption and statistical estimation precision is selected and highlighted in Figure 5-6 using a circle. Based on Table 5-5, the three optimum ALT plans regarding individual optimization criteria are selected and listed in Table 5-6 along with this overall optimum ALT plan.

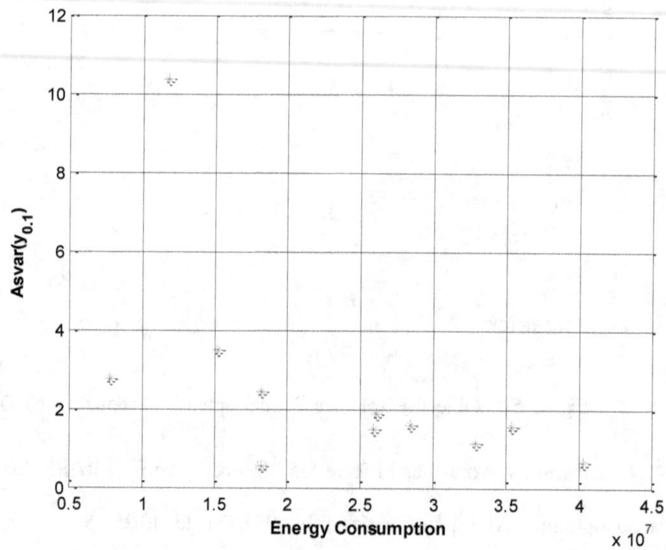

Figure 5-6: Objective values of feasible Pareto optimal solutions in a 2-D space

Table 5-6: Selected optimum ALT plans (Unadjustable PID)

Solution #	$\underline{X} = \left[Z_1, Z_2, \tau_1, \tau, F_p, F_d\right]$	$f_1(\underline{X})$	$f_2(\underline{X})$	$f_3(\underline{X})$	Note
32	[46, 98, 38.4, 48.6, 0.4195, 1043.16]	995,431	1388.139	1.8760	Minimal energy consumption
3	[98.3, 150, 354.0, 379.7, 1.1392, 3000.47]	18,225,132	0.5432	0.6275	Optimal estimate accuracy
2	[92, 147, 342.8, 367.2, 1.2276, 6388.69]	40,251,896	0.6420	0.0169	Minimal RMSE
3*	[98, 150, 354, 379.7, 1.1392, 3000.47]	18,225,132	0.5432	0.6275	Overall optimum ALT

As seen in Table 5-6, Solution #32 consumes the lease amount of energy but results in extremely large uncertainty in reliability estimation and unqualified tracking performance. Solution #2 sacrifices its energy and statistical efficiency to improve its tracking performance. Given trade-offs between energy consumption and estimation precision while considering the constraint on tracking performance, the overall optimum ALT plan designed using simultaneous optimization method is:

$$\underline{X}^* = \underline{X}^{(4)} = \left[Z_1^*, Z_2^*, \tau_1^*, \tau^*, F_p^*, F_d^*\right]$$
$$= [98, 150, 354, 379.7, 1.1392, 3000.47] \tag{5.32}$$

For this plan, $Asvar\left(\hat{y}_{0.1}(\hat{\underline{\Theta}}, Z_0, n_{total} = 50 \mid \underline{X}^*)\right) = 0.5432$ and the RMSE is equal to 0.6275 that is smaller than $1\,^{o}C$. The estimated energy consumption of this ALT plan is 18,225,132 units of energy.

5.3 Comparison of Results

Compared to the sequential optimization procedure, the simultaneous optimization procedure belongs to a large-scale optimization problem which supposes to generate better ALT plans regarding either statistical estimate accuracy or energy consumption, or even both. To verify this supposition, this section utilizes the sequential optimization method to design the statistically and energy efficient ALT plans considering the same setting as the case study conducted in Section 5.2 and then compares the resulting ALT plans with the overall ALT plans obtained using simultaneous optimization procedures.

5.3.1 Test Equipment with an Adjustable PID Controller

First, we assume the test equipment has an adjustable PID controller. Two levels of *constant-stress* loading (see Table 5-7) are utilized in sequential optimization process to design the statistically accurate ALT plan given in Problem *P1* with the decision variables: $\Omega^{(0)} = [Z_1^{(0)}, Z_2^{(0)}, n_1^{(0)}, n_2^{(0)}]$. Censoring time $\tau^{(0)}$ for the ALT plan with such stress loading is 200 hours. From the resulting ALT plan $[\Omega^{(0)*}, \tau^{(0)*}]$, we need to solve Problem *P2* to find an optimal equivalent simple SSALT plan with stress-loading Z_1^* (see Table 5-7), which has the same estimation precision as $[\Omega^{(0)*}, \tau^{(0)*}]$ while minimizing the total energy consumption for the entire test time τ^*. The decision variables of the desired statistically and energy efficient ALT plan are: $\Omega^{(1)} = Z_1, \tau$, and $[K_p, K_i, K_d]$ of the PID controller.

The mathematical formulation for this sequential ALT planning process consisting of two nonlinear programs is given by:

P1: **Statistically efficient ALT plan**

$$\Omega^{(0)*} = \underset{\Omega^{(0)}}{\arg\min} Asvar\left(\hat{y}_{0.1}(\hat{\Theta}, 25^\circ C, n_{total} = 50 \,|\, [\Omega^{(0)}, \tau^{(0)} = 200\text{hrs}])\right) \qquad (5.33)$$

Subject to $\quad n_1 + n_2 = n_{total} = 50, \ n_1, n_2 \in \{1,2,...\}$,

$$25^\circ C \le Z_i \le 150^\circ C, \qquad i = 1,2,$$

$$Z_2 \ge Z_1,$$

Take the solution $[\Omega^{(0)*}, \tau^{(0)}]$ as the input to $\rightarrow$ continued

$\rightarrow P2:$ **Desired statistically and energy efficient ALT plan**

$$[Z_1^*, \tau^*, K_p^*, K_i^*, K_d^*] = \underset{Z_1, \tau, K_p, K_i, K_d}{\arg\min} \ \mathbb{EN}(\tau, \{u(t;Z_1), 0 \le t \le \tau\}) = \int_0^\tau 1 + u^2(t;Z_1) \, dt \quad (5.34)$$

Subject to $\quad Asvar\left(\hat{y}_{0.1}(\hat{\Theta}, 25^\circ C, n_{total} = 50 \,|\, [Z_1, \tau])\right)$

$$= Asvar\left(\hat{y}_{0.1}(\hat{\Theta}, 25^\circ C \,|\, [\Omega^{(0)}, \tau^{(0)} = 200\text{hrs}])\right),$$

$$\left. \begin{array}{l} 25^\circ C \le Z_1 \le 150^\circ C, \\ 0 < \tau \le 200 \text{ hours}, \\ Z_2 \ge Z_1 \end{array} \right\} \text{same as those in } P1$$

$$\mathbb{L}_s\left(\tilde{Z}_1(t)\right) = \mathbb{L}_s\left(u(t;Z_1)\right) H(s), \quad \text{(equipment's response to input)}$$

$$RMSE = \sqrt{\frac{\sum_{j=1}^{N}(Z_1(t_j) - \tilde{Z}_1(t_j))^2}{N}} \le 0.5^\circ C, \quad 0 \le t_j \le \tau,$$

$$u(t;Z_1) = K_p e(t) + K_i \int_0^t e(v) \, dv + K_d [de(t)/dt],$$

$$0.1 < K_p \le 5,$$

$$0 < K_i \le 1.5,$$

$$0 < K_d \le 100,$$

When the *constant-stress* loadings are used in ALT experiments, the corresponding

pdf $f(t;Z_i)$ and cdf $F(t;Z_i)$ of failure times and reliability function $R(t;Z_i)$ under Z_i

can be expressed as (Liao and Elayed, 2010):

$$f(t;Z_i) = \frac{1}{\sigma t} \exp\left[\frac{\ln(t) - \mu_i}{\sigma} - \exp(\frac{\ln(t) - \mu_i}{\sigma}) \right], \qquad (5.35)$$

$$F(t; Z_i) = 1 - R(t; Z_i) = 1 - \exp\left[-\exp(\frac{\ln(t) - \mu_i}{\sigma})\right], \tag{5.36}$$

respectively. Here, $\mu_i = \theta_0 + \theta_1\left[1/(Z_0 + 273.16) - 1/(Z_i + 273.16)\right]$, $i = \{1, 2\}$.

The associated log-likelihood function is given by:

$$L(\underline{\Theta} \mid \text{data}) = \sum_{i=1}^{2} \sum_{j=1}^{n_i} \left[\delta_{ij} \ln\left(f(t_{ij}; Z_i)\right) + (1 - \delta_{ij}) \ln\left(R(\tau; Z_i)\right)\right], \tag{5.37}$$

where $\delta_{ij} = \{1, \text{ if } t_{ij} < \tau; 0, \text{ otherwise}\}$. Then, estimation precision of the ALT experiments interested in Problem $P1$ and Problem $P2$ can be represented by:

$$Asvar\left(\hat{y}_{0.1}(\hat{\underline{\Theta}}, 25^\circ C \mid [\Omega^{(0)}, \tau^{(0)}])\right) = \left[\frac{\partial \hat{y}_{0.1}(\hat{\underline{\Theta}}, 25^\circ C)}{\partial \hat{\underline{\Theta}}}\right]^T \hat{\Sigma}_{\hat{\underline{\Theta}}}^{[\Omega^{(0)}, \tau^{(0)}]} \left[\frac{\partial \hat{y}_{0.1}(\hat{\underline{\Theta}}, 25^\circ C)}{\partial \hat{\underline{\Theta}}}\right] \tag{5.38}$$

and

$$Asvar\left(\hat{y}_{0.1}(\hat{\underline{\Theta}}, 25^\circ C \mid [Z_1, \tau, n_{total}])\right) = \left[\frac{\partial \hat{y}_{0.1}(\hat{\underline{\Theta}}, 25^\circ C)}{\partial \hat{\underline{\Theta}}}\right]^T \hat{\Sigma}_{\hat{\underline{\Theta}}}^{[Z_1, \tau, n_{total}]} \left[\frac{\partial \hat{y}_{0.1}(\hat{\underline{\Theta}}, 25^\circ C)}{\partial \hat{\underline{\Theta}}}\right] \tag{5.39}$$

using

$$\left[\frac{\partial \hat{y}_{0.01}(\hat{\underline{\Theta}}, 25^\circ C)}{\partial \hat{\underline{\Theta}}}\right]^T = \left[1, 0, \Phi_{sev}^{-1}(0.1)\right] \tag{5.40}$$

and the corresponding variance-covariance matrices $\hat{\Sigma}_{\hat{\underline{\Theta}}}^{[\cdot]} = \left[E\left[-\frac{\partial^2 L(\underline{\Theta} \mid \text{data})}{\partial \underline{\Theta} \partial \underline{\Theta}^T}\right]\right]^{-1}$ evaluated at

$\hat{\underline{\Theta}}$. (See Appendix B and Appendix C for derivation process and the general expressions

of the terms in $E\left[-\frac{\partial^2 L(\underline{\Theta} \mid \text{data})}{\partial \underline{\Theta} \partial \underline{\Theta}^T}\right]$ for the ALT experiments using either the *constant-stress*

loadings or the *simple step-stress* loading).

Solve Problem $P1$ using Excel and get the traditional optimum ALT plan focusing

only on statistical efficiency: $Z_1^{(0)*} = 100$, $Z_2^{(0)*} = 150$, $n_1^{(0)*} = 35$, and $n_2^{(0)*} = 15$. To Solve

Problem $P2$, after obtaining the statistically accurate ALT plan $\Omega^{(0)*}$, we first solve the first equality considering all inequality constraints same as those in Problem $P1$ using the method proposed by Liao and Elsayed (2010) and obtain a set of simple SSALT plans. Each of these plans has different stress loading Z_1 and test time τ but they all can provide the same statistical estimation precision as the benchmark ALT obtained in Problem $P1$. Then an auto-tuning algorithm is executed for each generated ALT plan to select the optimal values of PID parameters which can make $\tilde{Z}_1$ follow Z_1 with RMSE $\leq$ $0.5^\circ C$. If feasible, the energy consumption associated with current ALT plan is calculated using the input signal produced by the corresponding optimal PID controller and this plan is then added in to the candidate ALT plan set; otherwise, this ALT plan is eliminated. By comparing the energy consumption of each candidate ALT plan, we can find the desired ALT plan $[Z_1^*, \tau^*]$ along with the controller parameters $[K_p^*, K_i^*, K_d^*]$ which consumes the least amount of energy.

Table 5-7 presents the optimal ALT plans obtained by solving Problem $P1$ and $P2$ and their corresponding values of energy usage. Compared to the benchmark ALT plan targeting only on the estimation precision, the desired ALT plan reduces the energy consumption by 28.1% while retaining the optimal estimation precision.

Table 5-7: Results for the sequential optimization (adjustable PID)

ALT plans	Stress loadings and other decision variables	Parameters of PID controller, control signal $u(t; Z_i)$, and total energy use
Benchmark statistically efficient ALT plan $Asvar\left(\hat{y}_{0.1}(25^{o}C)\right)$ $= 0.5997$	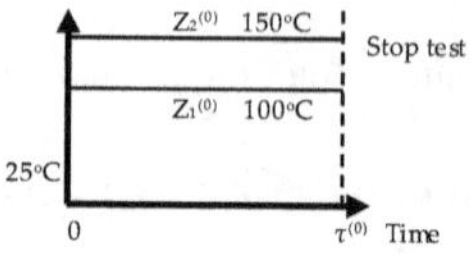 $n_1^{(0)*} = 35, n_2^{(0)*} = 15,$ $\tau^{(0)} = 200$ hours. **Two levels of constant-stress loading**	$K_p^* = 4.9904,$ $K_i^* = 0.75,$ $K_d^* = 0.01.$ 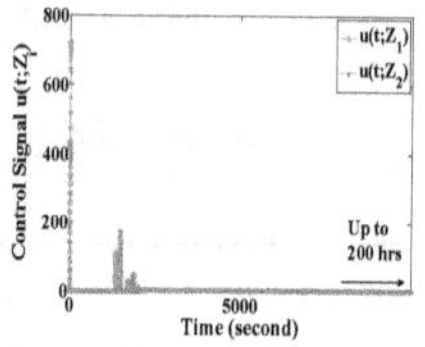 Total energy use $= 26,544,802$ units of energy
Desired statistically & energy efficient ALT plan $Asvar\left(\hat{y}_{0.1}(25^{o}C)\right)$ $= 0.5995$	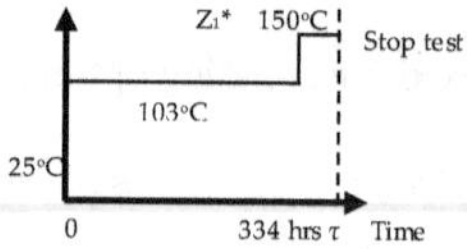 $n_{total} = 50,$ $\tau^* = 380$ hours. **One simple step-stress loading**	$K_p^* = 4.9904,$ $K_i^* = 0.75,$ $K_d^* = 87.3060.$ 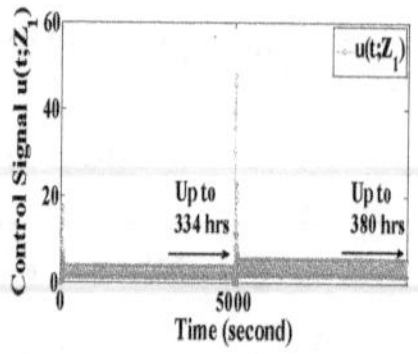 Total energy use $= 19,078,189$ units of energy

For ease of comparison, the two statistically and energy efficient ALT plans obtained using sequential and simultaneous optimization approaches are summarized in Table 5-8. Compared with the sequentially optimum ALT plan, the overall optimum ALT plan designed by simultaneous optimization method further reduces the energy consumption by 19.2% and improves estimation precision by 11.21%. Although the RMSE of this ALT plan is increased due to the significant improvements in the other two objectives, it is still within the acceptable range ($< 1\ ^{o}C$). In general, the overall optimum ALT plan obtained in Section 5.2.1 is better than the ALT plan designed using sequential optimization, which is consistent with the supposition discussed at the beginning of this section.

Table 5-8: The desired statistically and energy efficient ALT plans

ALT Plan $\left[Z_1, Z_2, \tau_1, \tau, K_p, K_i, K_d\right]$	Total Energy	$As\,\mathrm{var}(\hat{y}_{0.1})$	RMSE	Method
[103, 150, 334, 380.0, 4.9904, 0.75, 87.3060]	19,078,189	0.5995	0.1931	Sequential Optimization
[97, 150, 361.2, 380.0, 4.9986, 0.62, 7.6678]	15,415,517	0.5323	0.6578	Simultaneous Optimization

5.3.2 Test Equipment with an Unadjustable PID Controller

When the test equipment has an unadjustable PID controller with $K_p = 0.11$, $K_i = 0.001$ and $K_d = 0.01$, we need to introduce input-weighting PID controller to improve energy utilization efficiency and tracking performance of test equipment. Hence, in this case Problem $P2$ in Eq. (5.33) should be changed to:

$\rightarrow P2$: **Desired statistically and energy efficient ALT plan**

$$[Z_1^*, \tau^*, F_p^*, F_d^*] = \underset{Z_1, \tau, F_p, F_d}{\arg\min} \; \mathbb{EN}(\tau, \{u(t; Z_1), 0 \le t \le \tau\}) = \int_0^\tau 1 + u^2(t; Z_1)\, dt \tag{5.41}$$

$$\text{Subject to} \quad Asvar\left(\hat{y}_{0.1}(\hat{\underline{\Theta}}, 25^\circ C, n_{total} = 50\,|\,[Z_1, \tau])\right)$$

$$= Asvar\left(\hat{y}_{0.1}(\hat{\underline{\Theta}}, 25^\circ C\,|\,[\Omega^{(0)}, \tau^{(0)} = 200\mathrm{hrs}])\right),$$

$$RMSE = \sqrt{\frac{\sum_{j=1}^{N}(Z_1(t_j) - \bar{Z}_1(t_j))^2}{N}} \le 1^\circ C, \quad 0 \le t_j \le \tau,$$

$$25^\circ C \le Z_i \le 150^\circ C, i = \{1, 2\},$$

$$10 < \tau_1 \le 380 \text{ hours},$$

$$\tau_1 < \tau \le 380 \text{ hours},$$

$$\mathbb{L}_s\big(\tilde{Z}_1(t)\big) = \mathbb{L}_s\big(u(t;Z_1)\big)H(s), \quad \text{(equipment's response to input)}$$

$$u(t;Z_i) = K_p\Big[F_p Z_i(t) - \tilde{Z}_i(t)\Big] + K_i\int_0^t \Big[F_i Z_i(t) - \tilde{Z}_i(t)\Big]\,dv$$

$$+ K_d\Big[d\big(F_d Z_i(t) - \tilde{Z}_i(t)\big)/dt\Big],$$

$$0.05 < F_p \le 1.5,$$

$$1 < F_d \le 7000,$$

Given the traditional optimum ALT plan $\Omega^{(0)*}$ (see Table 5-7), Eq. (5.41) can be solved using the algorithms developed in Section 3.4.2. Table 5-9 summarizes the statistically efficient ALT plan obtained in Section 5.3.1 and the desired ALT plan obtained by solving Eq. (5.41). The associated controllers and the corresponding values of energy consumption are also listed in this table. As seen on Table 5-9, when PID controller is unadjustable, compared to the benchmark ALT plan, the desired ALT plan reduces the energy consumption by 26.6% while retaining the optimal estimation precision.

Table 5-9: Results for the statistically efficient ALT plan and the desired ALT plan (C2)

ALT plans	Stress loadings and other decision variables	Parameters of PID controller, control signal $u(t;Z_i)$, and total energy use
Benchmark statistically efficient ALT plan $Asvar\big(\hat{y}_{0.1}(25^{\circ}C)\big)$ $= 0.5997$	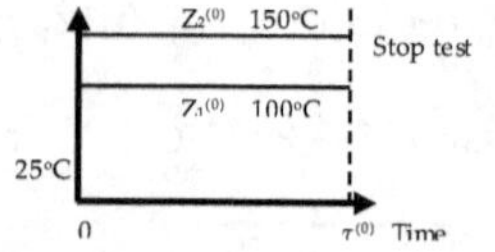 $n_1^{(0)*} = 35, n_2^{(0)*} = 15,$ $\tau^{(0)} = 200$ hours. **Two levels of constant-stress loading**	$K_p^* = 0.11,$ $K_i^* = 0.001,$ $K_d^* = 0.01.$ Total energy use $= 27,005,228$ units of energy 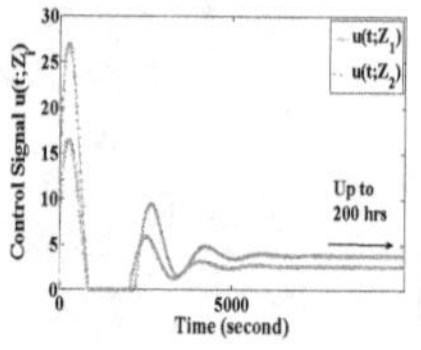

Desired statistically & energy efficient ALT plan $Asvar\left(\hat{y}_{0.1}(25^{\circ}C)\right)$ $= 0.5995$	One simple step-stress loading $n_{total} = 50,$ $\tau^{*} = 380$ hours.	$F_{p}^{*}=0.1001,$ $F_{d}^{*}=3000.$ Total energy use $= 19,833,851$ units of energy	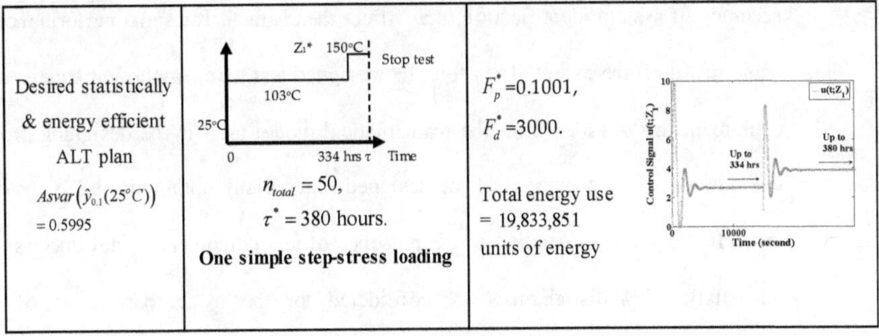

Table 5-8 summarizes two statistically and energy efficient ALT plans obtained using sequential and simultaneous optimization approaches in the case where the PID controller of test equipment is fixed by the manufacturer. Compared with the sequentially optimum ALT plan, the overall optimum ALT plan designed by simultaneous optimization method further reduces the energy consumption by 8.1%, improves estimation precision by 9.4% and reduces the RMSE by 26.7%. Similar to what we have observed in the previous section, the overall optimum ALT plan obtained in Section 5.2.2 is also superior to the ALT plan designed using sequential optimization.

Table 5-10: The desired statistically and energy efficient ALT plans

ALT Plan $[Z_1, Z_2, \tau_1, \tau, \underline{C}]$	Total Energy	$As\,\mathrm{var}(\hat{y}_{0.1})$	RMSE	Method
[103, 150, 334, 380.0, 0.1001, 3000]	19,833,851	0.5995	0.8564	Sequential Optimization
[98, 150, 354, 379.7, 1.1392, 3000.47]	18,225,132	0.5432	0.6275	Simultaneous Optimization

5.4 Sensitive Analysis

In practice, factors, such as the natural variation in the laboratory environment and the accuracy of system identification, may affect the characteristics and performance of test equipment to some extent. Therefore, the designed test plan may be implemented on test equipment that deviates from the mathematical model used in the designing process. In this section, the performance of the designed sequentially optimum ALT plans obtained in Section 5.1 will be studied in the presence of test equipment model uncertainties. In this study, ±5% disturbances are considered for the model parameters of the test equipment. Table 5-11 summarizes the performance of the controllers associated to the designed statistically and energy efficient ALT plans in both undisturbed and disturbed test systems. The optimal controllers designed for the disturbed system and their performance are also provided in Table 5-11.

As shown in Table 5-11, when the PID controller is adjustable (Case 1), and there is 5% disturbance in the system model, compared with the undisturbed system, the energy consumption decreases by 7.6% and the associated RMSE increases by 0.0009 $^\circ C$; when the system has -5% disturbance, the energy consumption increases by 8.8% and RMSE decreases by 0.001 $^\circ C$. The values of optimal PID control parameters remain the same when the system is subject to ±5% disturbances. When the PID controller is unadjustable, for the under-damped system (Case 2.1), the optimal input-weighting PID controller is quite robust to disturbances in system parameters, meanwhile the corresponding energy consumption decreases by 8.1% and the RMSE decreases by 4.0% for 5% disturbance; the -5% system parameter disturbance increases the energy consumption by 9.4% and there exists 3.0% increase in RMSE. For the over-damped system (Case 2.2), both types of system disturbances affect the design of input-weighting PID controllers. Compared to the undisturbed system, when the system has 5% disturbance, the energy consumption decreases by 8.0% and the associated RMSE decreases by 16.5%; when the system has -5% disturbance, the energy consumption increases by 9.3% and RMSE increases by 18.2%. In summary, the statistically and energy efficient ALT plans designed using sequential optimization approach are robust to system disturbances to some extent and are able to reduce the energy consumption by over 58% even in the presence of system disturbances.

Table 5-11: Sensitivity analyses on the performance of the desired ALT plans (uncertain equipment model)

$\sigma\%$	$H(S)$	Case 1			Case 2.1 (under-damped)			Case 2.2 (over-damped)		
		PID	Energy used	Tracking Error (RMSE)	IWPID	Energy used	Tracking Error (RMSE)	IWPID	Energy used	Tracking Error (RMSE)
0%	$\dfrac{38.8}{(2475S+1)}$	$K_p^*=4.9904$ $K_i^*=0.75$ $K_d^*=0.01$	**5,968,637**	0.1529	$F_p^*=0.10$ $F_d^*=6387.24$	**5,643,278**	0.0100	$F_p^*=1.17$ $F_d^*=5850.03$	**5,627,437**	**0.4723**
+5%	$\dfrac{40.74}{(2598.75S+1)}$	PID (used) $K_p^*=4.9904$ $K_i^*=0.75$ $K_d^*=0.01$	**5,513,995**	0.1538	IWPID (used) $F_p^*=0.10$ $F_d^*=6387.24$	**5,188,566**	0.0096	IWPID (used) $F_p^*=1.17$ $F_d^*=5850.03$	**5,176,460**	**0.3945**
		PID (needed) $K_p^{*'}=4.9904$ $K_i^{*'}=0.75$ $K_d^{*'}=0.01$	**5,513,995**	0.1538	IWPID (needed) $F_p^{*'}=0.10$ $F_d^{*'}=6387.24$	**5,188,566**	0.0096	IWPID (needed) $F_p^{*'}=1.16$ $F_d^{*'}=5793.67$	**5,174,228**	**0.4675**
-5%	$\dfrac{36.86}{(2351.25S+1)}$	PID (used) $K_p^*=4.9904$ $K_i^*=0.75$ $K_d^*=0.01$	**6,496,510**	0.1519	IWPID (used) $F_p^*=0.10$ $F_d^*=6387.24$	**6,171,564**	0.0103	IWPID (used) $F_p^*=1.17$ $F_d^*=5850.03$	**6,150,982**	**0.5581**
		PID (needed) $K_p^{*'}=4.9904$ $K_i^{*'}=0.75$ $K_d^{*'}=0.01$	**6,496,510**	0.1519	IWPID (needed) $F_p^{*'}=0.10$ $F_d^{*'}=6387.24$	**6,171,564**	0.0103	IWPID (needed) $F_p^{*'}=1.18$ $F_d^{*'}=5906.58$	**6,153,111**	**0.4855**

CHAPTER 6

CONCLUSION AND FUTURE WORK

6.1 Conclusions

A methodology for planning statistically and energy efficient ALT experiments has been investigated in this dissertation. To handle such unique experimental design problems, a new double-loop optimization framework is proposed. Unlike most optimum design of ALT plans, the new methodology utilizes the physical model of test equipment in computer simulation to design a capable controller for the equipment and to evaluate the energy consumption of the candidate experimental designs. Besides the energy consumption, the tracking performance of test equipment is also taken into account during controller design to create an accurate test environment. Based on the proposed framework, two optimization procedures have been developed in this dissertation: 1) sequential optimization approach which optimizes statistical efficiency of ALT followed by energy minimization through the optimum design of control strategy for the test equipment; 2) a fully integrated approach which can optimize the estimation precision, energy usage, and tracking performance of ALT experiment simultaneously.

A set of computational tools has been developed to facilitate the design of statistically and energy efficient ALT experiments for test equipment with adjustable PID controllers. In addition, another set of experimental design tools based on the input-weighting method has also been developed to deal with situations where the given PID controller is hardwired and/or its adjustment is prohibited. The numerical examples demonstrated that the resulting statistically and energy efficient ALT plans have significant potential for reducing the total energy consumption without sacrificing the estimation precision in ALT. Sensitivity analysis results also showed the robustness of the resulting sequentially optimum ALT plans against the uncertainty in the test equipment model. It is worth mentioning that although the developed optimization algorithms in this dissertation are based on PID controllers, the concept can be applied to other types of control strategies, such as on/off control and state-space control.

It is expected that many industries, such as the automotive industry and microelectronics industry, can benefit from the reduction in total energy consumption, uncertainty in product reliability, and total costs of developing a reliable product.

6.2 Future Work

To improve the proposed new ALT design methodology, several areas are in need of further exploration:

- In this dissertation, RMSE is used to quantify the tracking performance of test equipment. Although RMSE gives a relatively higher weight to large errors between the desired stress profile and the actual test condition created by the test equipment, it still cannot fully represent the magnitude of overshoots caused by the adopted control strategies. Since significant overshoot ($> 10\ ^oC$) in the created test condition may introduce new failure modes that would not occur under normal operation conditions, a new constraint could be introduced in the ALT design processes to ensure the absolute values of error fall within an acceptable range.

- The simultaneous optimization procedure developed in this dissertation has a high computational complexity of $O(MN^2)$, where M is the number of objectives and N is the population size (Deb and Goel, 2001). Since this design procedure requires simulating the test equipment in estimating the energy consumption of ALT experiment, it normally takes a long time to generate Pareto optimal solutions. In order to reduce the computation time of such ALT designs, the algorithms with lower computational complexity should be investigated in the future.